PEARSON EDEXCEL INTERNATIONAL A LEVEL
MECHANICS 1
Student Book

Series Editors: Joe Skrakowski and Harry Smith

Authors: Greg Attwood, Jack Barraclough, Ian Bettison, Linnet Bruce, Alan Clegg, Gill Dyer, Jane Dyer, Keith Gallick, Susan Hooker, Michael Jennings, Mohammed Ladak, Jean Littlewood, Alistair Macpherson, Bronwen Moran, James Nicholson, Su Nicholson, Diane Oliver, Laurence Pateman, Keith Pledger, Joe Skrakowsi, Harry Smith, Geoff Staley, Robert Ward-Penny, Jack Williams, Dave Wilkins

2019/M1.5

Published by Pearson Education Limited, 80 Strand, London, WC2R 0RL.

www.pearsonglobalschools.com

Copies of official specifications for all Pearson qualifications may be found on the website: https://qualifications.pearson.com

Text © Pearson Education Limited 2018
Edited by Eric Pradel
Designed by © Pearson Education Limited 2018
Typeset by © Tech-Set Ltd, Gateshead, UK
Original illustrations © Pearson Education Limited 2018
Illustrated by © Tech-Set Ltd, Gateshead, UK
Cover design by © Pearson Education Limited 2018

The rights of Greg Attwood, Jack Barraclough, Ian Bettison, Linnet Bruce, Alan Clegg, Gill Dyer, Jane Dyer, Keith Gallick, Susan Hooker, Michael Jennings, Mohammed Ladak, Jean Littlewood, Alistair Macpherson, Bronwen Moran, James Nicholson, Su Nicholson, Diane Oliver, Laurence Pateman, Keith Pledger, Joe Skrakowsi, Harry Smith, Geoff Staley, Robert Ward-Penny, Jack Williams and Dave Wilkins to be identified as the authors of this work have been asserted by them in accordance with the Copyright, Designs and Patents Act 1988.

First published 2018

21 20 19
10 9 8 7 6 5 4 3 2

British Library Cataloguing in Publication Data
A catalogue record for this book is available from the British Library

ISBN 978 1 292 24467 9

Copyright notice
All rights reserved. No part of this publication may be reproduced in any form or by any means (including photocopying or storing it in any medium by electronic means and whether or not transiently or incidentally to some other use of this publication) without the written permission of the copyright owner, except in accordance with the provisions of the Copyright, Designs and Patents Act 1988 or under the terms of a licence issued by the Copyright Licensing Agency, 5th Floor, Shackleton House, 4 Battlebridge Lane, London, SE1 2HX (www.cla.co.uk). Applications for the copyright owner's written permission should be addressed to the publisher.

Printed in Slovakia by Neografia

Picture Credits
The authors and publisher would like to thank the following individuals and organisations for permission to reproduce photographs:

(Key: b-bottom; c-centre; l-left; r-right; t-top)

Fotolia.com: Arousa 54; **Getty Images:** Jeff Schultz 39; **Shutterstock.com:** Algonga 7, Carlos Castilla 101, Carlos. E. Santa Maria 1, Joggie Botma 7, Lane V. Erickson 84, mbolina 112, Sinesp 10, **123RF.com:** dmitrimaruta 129

Cover images: *Front:* **Getty Images:** Werner Van Steen
Inside front cover: **Shutterstock.com:** Dmitry Lobanov

All other images © Pearson Education
All artwork © Pearson Education

Endorsement Statement
In order to ensure that this resource offers high-quality support for the associated Pearson qualification, it has been through a review process by the awarding body. This process confirms that this resource fully covers the teaching and learning content of the specification or part of a specification at which it is aimed. It also confirms that it demonstrates an appropriate balance between the development of subject skills, knowledge and understanding, in addition to preparation for assessment.

Endorsement does not cover any guidance on assessment activities or processes (e.g. practice questions or advice on how to answer assessment questions) included in the resource, nor does it prescribe any particular approach to the teaching or delivery of a related course.

While the publishers have made every attempt to ensure that advice on the qualification and its assessment is accurate, the official specification and associated assessment guidance materials are the only authoritative source of information and should always be referred to for definitive guidance.

Pearson examiners have not contributed to any sections in this resource relevant to examination papers for which they have responsibility.

Examiners will not use endorsed resources as a source of material for any assessment set by Pearson. Endorsement of a resource does not mean that the resource is required to achieve this Pearson qualification, nor does it mean that it is the only suitable material available to support the qualification, and any resource lists produced by the awarding body shall include this and other appropriate resources.

CONTENTS

COURSE STRUCTURE	iv
ABOUT THIS BOOK	vi
QUALIFICATION AND ASSESSMENT OVERVIEW	viii
EXTRA ONLINE CONTENT	x
1 MATHEMATICAL MODELS IN MECHANICS	1
2 CONSTANT ACCELERATION	10
3 VECTORS IN MECHANICS	39
4 DYNAMICS OF A PARTICLE MOVING IN A STRAIGHT LINE	54
REVIEW EXERCISE 1	79
5 FORCES AND FRICTION	84
6 MOMENTUM AND IMPULSE	101
7 STATICS OF A PARTICLE	112
8 MOMENTS	129
REVIEW EXERCISE 2	146
EXAM PRACTICE	150
GLOSSARY	153
ANSWERS	155
INDEX	166

CHAPTER 1 MATHEMATICAL MODELS IN MECHANICS — 1
- 1.1 CONSTRUCTING A MODEL — 2
- 1.2 MODELLING ASSUMPTIONS — 4
- 1.3 QUANTITIES AND UNITS — 6
- CHAPTER REVIEW 1 — 8

CHAPTER 2 CONSTANT ACCELERATION — 10
- 2.1 DISPLACEMENT–TIME GRAPHS — 11
- 2.2 VELOCITY–TIME GRAPHS — 13
- 2.3 ACCELERATION-TIME GRAPHS — 17
- 2.4 CONSTANT ACCELERATION FORMULAE 1 — 19
- 2.5 CONSTANT ACCELERATION FORMULAE 2 — 24
- 2.6 VERTICAL MOTION UNDER GRAVITY — 29
- CHAPTER REVIEW 2 — 35

CHAPTER 3 VECTORS IN MECHANICS — 39
- 3.1 WORKING WITH VECTORS — 40
- 3.2 SOLVING PROBLEMS WITH VECTORS WRITTEN USING I AND J NOTATION — 42
- 3.3 THE VELOCITY OF A PARTICLE AS A VECTOR — 45
- 3.4 SOLVING PROBLEMS INVOLVING VELOCITY AND TIME USING VECTORS — 46
- CHAPTER REVIEW 3 — 50

CHAPTER 4 DYNAMICS OF A PARTICLE MOVING IN A STRAIGHT LINE — 54
- 4.1 FORCE DIAGRAMS — 55
- 4.2 FORCES AS VECTORS — 58
- 4.3 FORCES AND ACCELERATION — 60
- 4.4 MOTION IN TWO DIMENSIONS — 64
- 4.5 CONNECTED PARTICLES — 67
- 4.6 PULLEYS — 71
- CHAPTER REVIEW 4 — 75

REVIEW EXERCISE 1 — 79

CHAPTER 5 FORCES AND FRICTION — 84
- 5.1 RESOLVING FORCES — 85
- 5.2 INCLINED PLANES — 90
- 5.3 FRICTION — 94
- CHAPTER REVIEW 5 — 99

CHAPTER 6 MOMENTUM AND IMPULSE — 101
- 6.1 MOMENTUM IN ONE DIMENSION — 102
- 6.2 CONSERVATION OF MOMENTUM — 104
- CHAPTER REVIEW 6 — 109

CHAPTER 7 STATICS OF A PARTICLE — 112

7.1 STATIC PARTICLES — 113
7.2 MODELLING WITH STATICS — 117
7.3 FRICTION AND STATIC PARTICLES — 121
CHAPTER REVIEW 7 — 126

CHAPTER 8 MOMENTS — 129

8.1 MOMENTS — 130
8.2 RESULTANT MOMENTS — 132
8.3 EQUILIBRIUM — 133
8.4 CENTRES OF MASS — 136
8.5 TILTING — 139
CHAPTER REVIEW 8 — 141

REVIEW EXERCISE 2 — 146

EXAM PRACTICE — 150

GLOSSARY — 153

ANSWERS — 155

INDEX — 166

ABOUT THIS BOOK

The following three overarching themes have been fully integrated throughout the Pearson Edexcel International Advanced Level in Mathematics series, so they can be applied alongside your learning and practice.

1. Mathematical argument, language and proof

- Rigorous and consistent approach throughout
- Notation boxes explain key mathematical language and symbols
- Opportunities to critique arguments and justify methods

2. Mathematical problem-solving

The Mathematical Problem-Solving Cycle

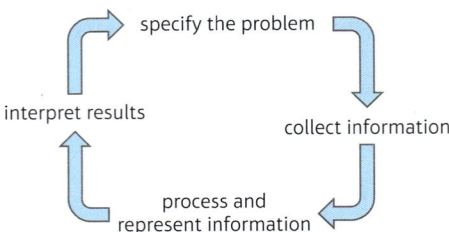

- Hundreds of problem-solving questions, fully integrated into the main exercises
- Problem-solving boxes provide tips and strategies
- Structured and unstructured questions to build confidence
- Challenge questions provide extra stretch

3. Mathematical modelling

- Dedicated modelling sections in relevant topics provide plenty of practice where you need it
- Examples and exercises include qualitative questions that allow you to interpret answers in the context of the model

Finding your way around the book

Access an online digital edition using the code at the front of the book.

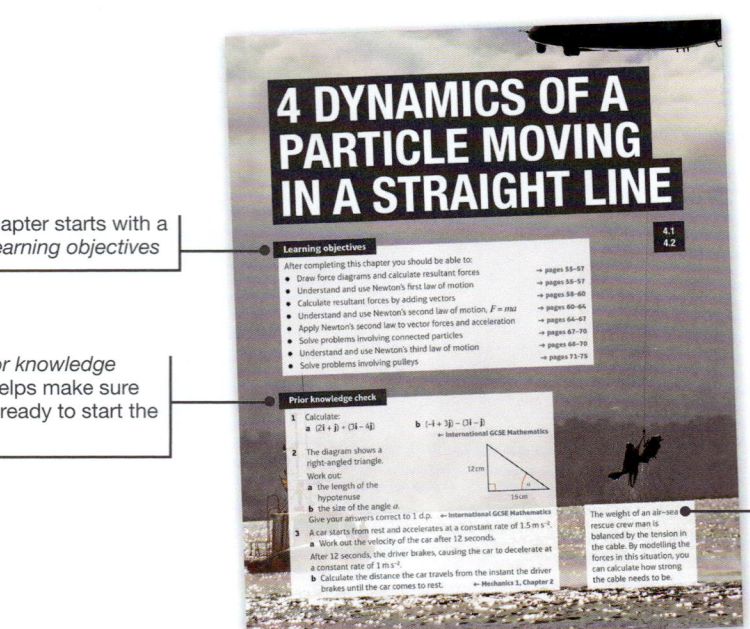

Each chapter starts with a list of *Learning objectives*

The *Prior knowledge check* helps make sure you are ready to start the chapter

The real world applications of the maths you are about to learn are highlighted at the start of the chapter

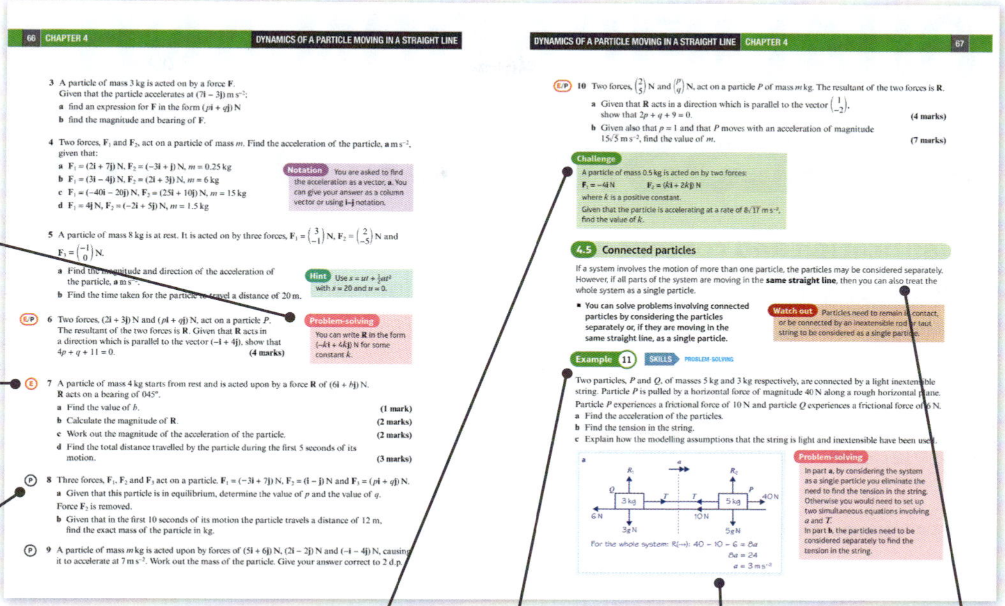

ABOUT THIS BOOK

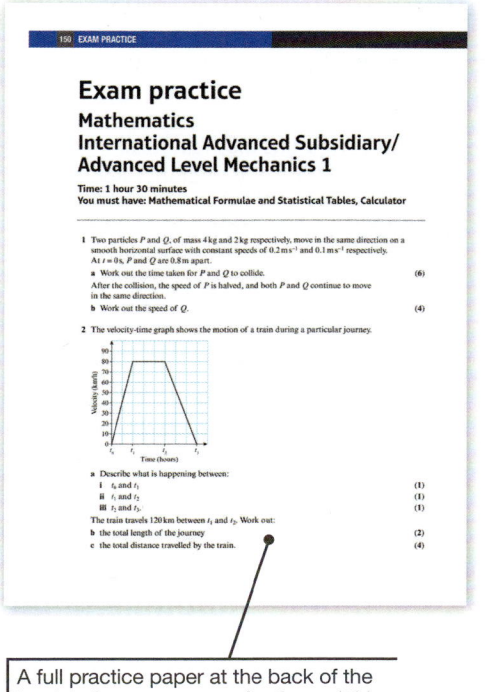

Problem-solving boxes provide hints, tips and strategies, and *Watch out* boxes highlight areas where students often lose marks in their exams

Exercises are packed with exam-style questions to ensure you are ready for the exams

Exercise questions are carefully graded so they increase in difficulty and gradually bring you up to exam standard

Exam-style questions are flagged with Ⓔ

Problem-solving questions are flagged with Ⓟ

Challenge boxes give you a chance to tackle some more difficult questions

Transferable skills are signposted where they naturally occur in the exercises and examples.

Each section begins with an explanation and key learning points

Step-by-step worked examples focus on the key types of questions you'll need to tackle

Each chapter ends with a *Chapter review* and a *Summary of key points*

After every few chapters, a *Review exercise* helps you consolidate your learning with lots of exam-style questions

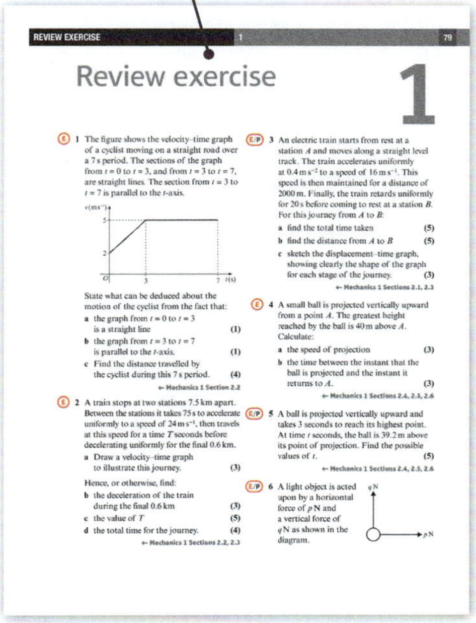

A full practice paper at the back of the book helps you prepare for the real thing

QUALIFICATION AND ASSESSMENT OVERVIEW

Qualification and content overview

Mechanics 1 (**M1**) is an **optional** unit in the following qualifications:

International Advanced Subsidiary in Mathematics
International Advanced Subsidiary in Further Mathematics
International Advanced Level in Mathematics
International Advanced Level in Further Mathematics

Assessment overview

The following table gives an overview of the assessment for this unit.

We recommend that you study this information closely to help ensure that you are fully prepared for this course and know exactly what to expect in the assessment.

Unit	Percentage	Mark	Time	Availability
M1: Mechanics 1	$33\frac{1}{3}$ % of IAS	75	1 hour 30 mins	January, June and October
Paper code WME01/01	$16\frac{2}{3}$ % of IAL			First assessment June 2019

IAS – International Advanced Subsidiary, IAL – International Advanced A Level

Assessment objectives and weightings

		Minimum weighting in IAS and IAL
AO1	Recall, select and use their knowledge of mathematical facts, concepts and techniques in a variety of contexts.	30%
AO2	Construct rigorous mathematical arguments and proofs through use of precise statements, logical deduction and inference and by the manipulation of mathematical expressions, including the construction of extended arguments for handling substantial problems presented in unstructured form.	30%
AO3	Recall, select and use their knowledge of standard mathematical models to represent situations in the real world; recognise and understand given representations involving standard models; present and interpret results from such models in terms of the original situation, including discussion of the assumptions made and refinement of such models.	10%
AO4	Comprehend translations of common realistic contexts into mathematics; use the results of calculations to make predictions, or comment on the context; and, where appropriate, read critically and comprehend longer mathematical arguments or examples of applications.	5%
AO5	Use contemporary calculator technology and other permitted resources (such as formulae booklets or statistical tables) accurately and efficiently; understand when not to use such technology, and its limitations. Give answers to appropriate accuracy.	5%

Relationship of assessment objectives to units

M1	Assessment objective				
	AO1	AO2	AO3	AO4	AO5
Marks out of 75	20–25	20–25	15–20	6–11	4–9
%	$26\frac{2}{3}–33\frac{1}{3}$	$26\frac{2}{3}–33\frac{1}{3}$	$20–26\frac{2}{3}$	$8–14\frac{2}{3}$	$5\frac{1}{3}–12$

Calculators

Students may use a calculator in assessments for these qualifications. Centres are responsible for making sure that calculators used by their students meet the requirements given in the table below.

Students are expected to have available a calculator with at least the following keys: $+$, $-$, $\times$, $\div$, π, x^2, $\sqrt{x}$, $\frac{1}{x}$, x^y, $\ln x$, e^x, $x!$, sine, cosine and tangent and their inverses in degrees and decimals of a degree, and in radians; memory.

Prohibitions

Calculators with any of the following facilities are prohibited in all examinations:

- databanks
- retrieval of text or formulae
- built-in symbolic algebra manipulations
- symbolic differentiation and/or integration
- language translators
- communication with other machines or the internet

Extra online content

Whenever you see an *Online* box, it means that there is extra online content available to support you.

SolutionBank

SolutionBank provides worked solutions for questions in the book.
Download the solutions as a PDF or quickly find the solution you need online.

Use of technology

Explore topics in more detail, visualise problems and consolidate your understanding. Use pre-made GeoGebra activities or Casio resources for a graphic calculator.

Online Find the point of intersection graphically using technology.

GeoGebra-powered interactives

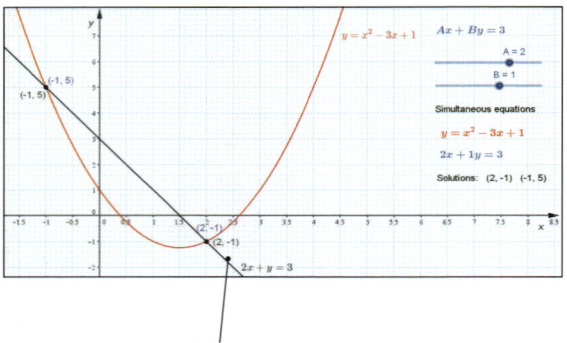

Interact with the maths you are learning using GeoGebra's easy-to-use tools

Graphic calculator interactives

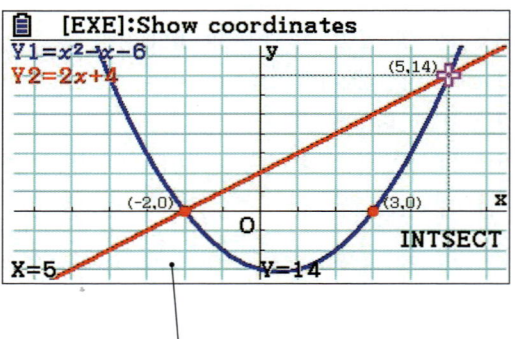

Explore the maths you are learning and gain confidence in using a graphic calculator

Calculator tutorials

Our helpful video tutorials will guide you through how to use your calculator in the exams. They cover both Casio's scientific and colour graphic calculators.

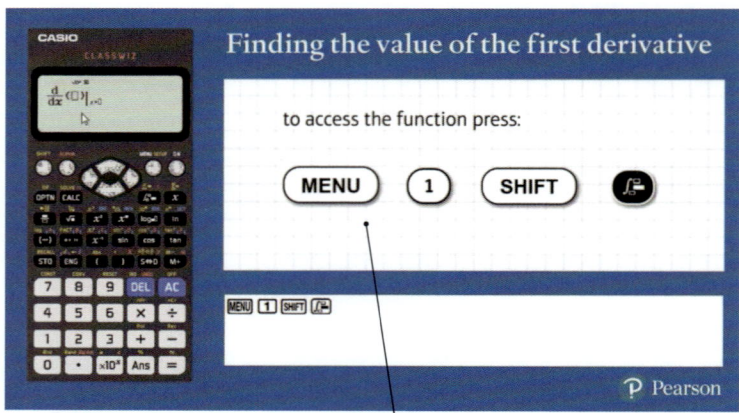

Online Work out each coefficient quickly using the ^nC_r and power functions on your calculator.

Step-by-step guide with audio instructions on exactly which buttons to press and what should appear on your calculator's screen

1 MATHEMATICAL MODELS IN MECHANICS

Learning objectives

After completing this chapter you should be able to:
- Understand how the concept of a mathematical model applies to mechanics → pages 2–3
- Understand and be able to apply some of the common assumptions used in mechanical models → pages 4–5
- Know SI units for quantities and derived quantities used in mechanics → pages 6–8

Prior knowledge check

Give your answers correct to 3 significant figures (s.f.) where appropriate.

1 Solve these equations:
   a $5x^2 - 21x + 4 = 0$
   b $6x^2 + 5x = 21$
   c $3x^2 - 5x - 4 = 0$
   d $8x^2 - 18 = 0$
   ← International GCSE Mathematics

2 Find the value of x and y in these right-angled triangles:
   a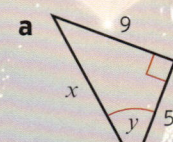
   b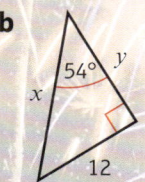
   ← International GCSE Mathematics

3 Convert:
   a 30 km h^{-1} to cm s^{-1}
   b 5 g cm^{-3} to kg m^{-3}
   ← International GCSE Mathematics

4 Write in standard form:
   a 7 650 000
   b 0.003 806
   ← International GCSE Mathematics

Mathematical models can be used to find solutions to **real-world problems** in many everyday situations. If you model a firework as a particle you can ignore the effects of wind and air resistance.

1.1 Constructing a model

Mechanics deals with **motion** and the action of **forces** on objects. Mathematical **models** can be constructed to simulate real-life situations (i.e. using models to create conditions that exist in real life, in order to study those conditions). However, in many cases it is necessary to simplify a problem by making one or more **assumptions**. This allows you to describe the problem using equations or graphs in order to solve it.

The solution to a mathematical model needs to be interpreted in the context of the original problem. It is possible that your model may need to be refined (improved with small changes) and your assumptions reconsidered.

This flow chart summarises the mathematical modelling process:

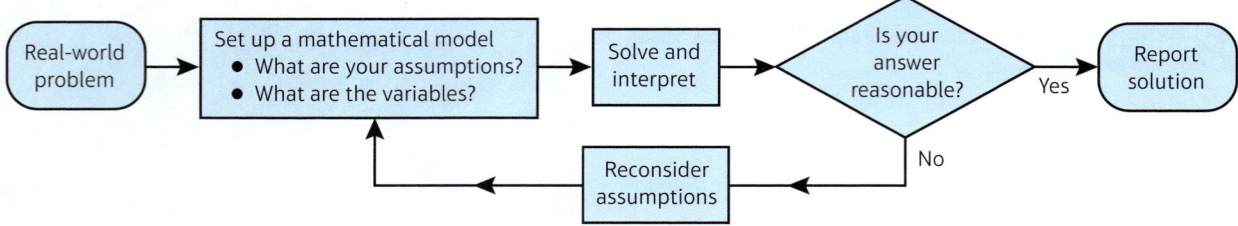

Example 1 — SKILLS: PROBLEM-SOLVING

The motion of a basketball as it leaves a player's hand and passes through the net can be modelled using the equation $h = 2 + 1.1x - 0.1x^2$, where h m is the height of the basketball above the ground and x m is the horizontal distance travelled.

a Find the height of the basketball:
  i when it is released
  ii at a horizontal distance of 0.5 m.
b Use the model to predict the height of the basketball when it is at a horizontal distance of 15 m from the player.
c Comment on the **validity** of this prediction.

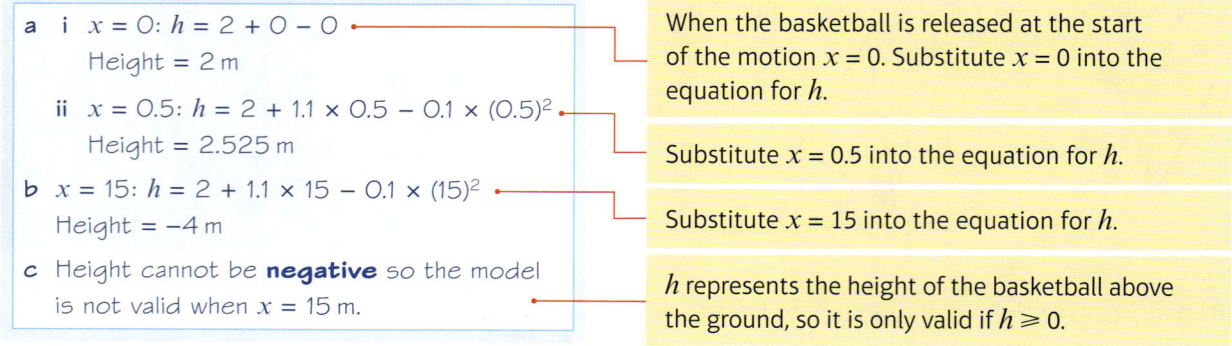

a i $x = 0$: $h = 2 + 0 - 0$
  Height = 2 m

  When the basketball is released at the start of the motion $x = 0$. Substitute $x = 0$ into the equation for h.

  ii $x = 0.5$: $h = 2 + 1.1 \times 0.5 - 0.1 \times (0.5)^2$
  Height = 2.525 m

  Substitute $x = 0.5$ into the equation for h.

b $x = 15$: $h = 2 + 1.1 \times 15 - 0.1 \times (15)^2$
  Height = −4 m

  Substitute $x = 15$ into the equation for h.

c Height cannot be **negative** so the model is not valid when $x = 15$ m.

h represents the height of the basketball above the ground, so it is only valid if $h \geqslant 0$.

Exercise 1A SKILLS PROBLEM-SOLVING

1. The motion of a golf ball after it is struck by a golfer can be modelled using the equation $h = 0.36x - 0.003x^2$, where h m is the height of the golf ball above the ground and x m is the horizontal distance travelled.
   a Find the height of the golf ball when it is:
      i struck
      ii at a horizontal distance of 100 m.
   b Use the model to predict the height of the golf ball when it is at a horizontal distance of 200 m from the golfer.
   c Comment on the validity of this prediction.

2. A stone is thrown into the sea from the top of a cliff. The height of the stone above sea level, h m at time t s after it is thrown can be modelled by the equation $h = -5t^2 + 15t + 90$.
   a Write down the height of the cliff above sea level.
   b Find the height of the stone:
      i when $t = 3$
      ii when $t = 5$.
   c Use the model to predict the height of the stone after 20 seconds.
   d Comment on the validity of this prediction.

(P) 3. The motion of a basketball as it leaves a player's hand and passes through the net is modelled using the equation $h = 2 + 1.1x - 0.1x^2$, where h m is the height of the basketball above the ground and x m is the horizontal distance travelled.
   a Find the two values of x for which the basketball is exactly 4 m above the ground.
   This model is valid for $0 \leqslant x \leqslant k$, where k m is the horizontal distance of the net from the player.
   b Given that the height of the net is 3 m, find the value of k.
   c Explain why the model is not valid for $x > k$.

(P) 4. A car accelerates from rest to 60 km h^{-1} in 10 seconds. A quadratic equation of the form $d = kt^2$ can be used to model the distance travelled, d metres in time t seconds.

   Problem-solving
   Use the information given to work out the value of k.

   a Given that when $t = 1$ second the distance travelled by the car is 13.2 metres, use the model to find the distance travelled when the car reaches 60 km h^{-1}.
   b Write down the range of values of t for which the model is valid.

(P) 5. The model for the motion of a golf ball given in question 1 is valid only when h is **positive**. Find the range of values of x for which the model is valid.

(P) 6. The model for the height of the stone above sea level given in question 2 is valid only from the time the stone is thrown until the time it enters the sea. Find the range of values of t for which the model is valid.

1.2 Modelling assumptions

Modelling assumptions can simplify a problem and allow you to analyse a real-life situation using known mathematical techniques. You need to understand the significance of different modelling assumptions and how they affect the calculations in a particular problem.

Watch out
Modelling assumptions can affect the validity of a model. For example, when modelling the landing of an aeroplane flight, it would not be appropriate to ignore the effects of wind and air resistance.

This table shows some common models and modelling assumptions that you need to know.

Model	Modelling assumptions
Particle – Dimensions of the object are **negligible**.	• **mass** of the object is concentrated at a single point • rotational forces (i.e. moving around a central fixed point) and air **resistance** can be ignored
Rod – All dimensions but one are negligible, like a pole or a beam.	• mass is concentrated along a line • no thickness • **rigid** (does not bend or buckle)
Lamina – Object with area but negligible thickness, like a sheet of paper.	• mass is distributed across a flat surface
Uniform body – Mass is distributed evenly.	• mass of the object is concentrated at a single point at the geometric centre of the **body** – the **centre of mass**
Light object – Mass of the object is small compared to other masses, like a string or a pulley.	• treat object as having zero mass • **tension** the same at both ends of a light string
Inextensible string – A string that does not stretch under load.	• **acceleration** is the same in objects connected by a **taut** inextensible string
Smooth surface – a surface on which it can be assumed there is no friction.	• assume that there is no **friction** between the surface and any object on it
Rough surface – a surface on which there is friction.	• objects in contact with the surface experience a frictional force if they are moving, or are acted on by a force
Wire – Rigid thin length of metal.	• treated as one-dimensional
Smooth and light pulley – All pulleys you consider will be smooth and light.	• pulley has no mass • tension is the same on either side of the pulley
Bead – Particle with a hole in it for threading on a wire or string (i.e. passing the wire or string through the hole).	• a smooth bead moves freely along a wire or string • for a smooth bead, tension is the same on either side of the bead
Peg – A support from which a body can be suspended or rested.	• dimensionless and fixed • can be rough or smooth as specified in the question
Air resistance – Resistance experienced as an object moves through the air.	• usually modelled as being negligible
Gravity – Force of attraction between all objects. Acceleration due to gravity is denoted by g. $g = 9.8 \text{ m s}^{-2}$	• assume all objects with mass are attracted toward the Earth • acceleration due to Earth's gravity is uniform (i.e. the same in all parts, at all times) and acts vertically downward • g is **constant** and is taken as 9.8 m s^{-2}, unless otherwise stated in the question

MATHEMATICAL MODELS IN MECHANICS — CHAPTER 1

Example 2 — SKILLS ANALYSIS

A mass is attached to a length of string which is fixed to the ceiling.

The mass is drawn to one side with the string stretched tightly and allowed to swing.

State the effect of the following assumptions on any calculations made using this model.

a The string is **light and inextensible** (unable to be stretched further).

b The mass is modelled as a particle.

> a Ignore the mass of the string and any stretching effect caused by the mass.
> b Ignore the rotational effect of any external forces that are acting on it, and the effects of air resistance.

Exercise 1B — SKILLS ANALYSIS

1 A football is kicked by the goalkeeper from one end of the football pitch.
   State the effect of the following assumptions on any calculations made using this model.
   a The football is modelled as a particle.
   b Air resistance is negligible.

2 An ice hockey puck is hit and slides across the ice.
   State the effect of the following assumptions on any calculations made using this model.
   a The ice hockey puck is modelled as a particle.
   b The ice is smooth.

3 A parachutist wants to model her descent from an aeroplane to the ground. She models herself and her parachute as particles connected by a light inextensible string. Explain why this may not be a suitable modelling assumption for this situation.

4 A fishing rod manufacturer constructs a mathematical model to predict the behaviour of a particular fishing rod. The fishing rod is modelled as a light rod.
   a Describe the effects of this modelling assumption.
   b Comment on its validity in this situation.

5 Make a list of the assumptions you might make to create simple models of the following:
   a the motion of a golf ball after it is hit
   b the motion of a child on a sledge going down a snow-covered hill
   c the motion of two objects of different masses connected by a string that passes over a pulley
   d the motion of a suitcase on wheels being pulled along a path by its handle.

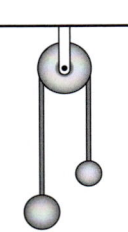

1.3 Quantities and units

The International System of Units, (abbreviated SI from the French, Système international d'unités) is the modern form of the metric system. These **base SI units** are most commonly used in mechanics:

Quantity	Unit	Symbol
Mass	kilogram	kg
Length/**displacement**	metre	m
Time	second	s

Watch out
A common misunderstanding is that kilograms measure **weight**, not mass. However, **weight** is a **force** which is measured in **newtons** (**N**).

These **derived** units are compound units built from the base units:

Quantity	Unit	Symbol
Speed/**velocity**	metres per second	m s^{-1}
Acceleration	metres per second per second	m s^{-2}
Weight/force	newton	N (= kg m s^{-2})

Example 3 — SKILLS: REASONING/ARGUMENTATION

Write the following quantities in SI units.

a 4 km **b** 0.32 grams **c** 5.1×10^6 km h^{-1}

a 4 km = 4 × 1000 = 4000 m
→ The SI unit of length is the metre; 1 km = 1000 m.

b 0.32 g = 0.32 ÷ 1000 = 3.2×10^{-4} kg
→ The SI unit of mass is the kg; 1 kg = 1000 g. The answer is given in standard form.

c 5.1×10^6 km h^{-1} = $5.1 \times 10^6 \times 1000$
 = 5.1×10^9 m h^{-1}
→ The SI unit of speed is m s^{-1}. Convert from km h^{-1} to m h^{-1} by multiplying by 1000.

$5.1 \times 10^9 \div (60 \times 60) = 1.42 \times 10^6$ m s^{-1}
→ Convert from m h^{-1} to m s^{-1} by dividing by 60 × 60. The answer is given in standard form to 3 s.f.

You will encounter a variety of forces in mechanics. These **force diagrams** show some of the most common forces.

- The **weight** (or gravitational force) of an object acts vertically downward.
- The **normal reaction** is the force which acts perpendicular (i.e. at a 90° angle to it) to a surface when an object is in contact with the surface. In this example the normal reaction is due to the weight of the book resting on the surface of the table.

R — Normal reaction exerted on the book (i.e. applied to it) by the table.

W — Force exerted on the table by the book. Both forces have the same **magnitude**.

- **Friction** is a force which opposes the motion between two rough surfaces.
- If an object is being pulled along by a string, the force acting on the object is called the **tension** in the string.
- If an object is being pushed along using a light rod, the force acting on the object is called the **thrust** or **compression** in the rod.

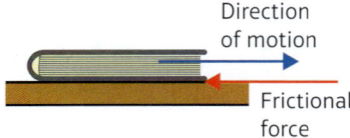
Direction of motion / Frictional force

Tension in string

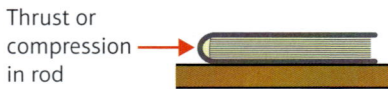
Thrust or compression in rod

- **Buoyancy** is the upward force on a body that allows it to float or rise when submerged (i.e. underneath the surface) in a liquid. In this example buoyancy acts to keep the boat afloat in the water.

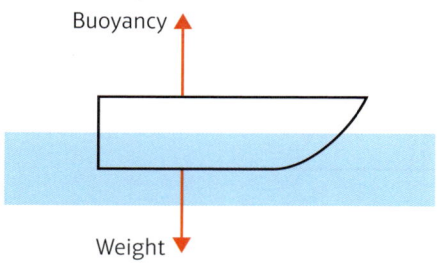

- **Air resistance** opposes motion. In this example the weight of the parachutist acts vertically downward and the air resistance acts vertically upward.

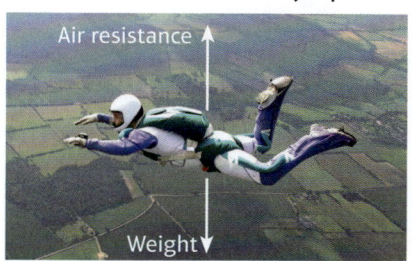

Example 4

The force diagram shows an aircraft in flight. Write down the names of the four forces shown on the diagram.

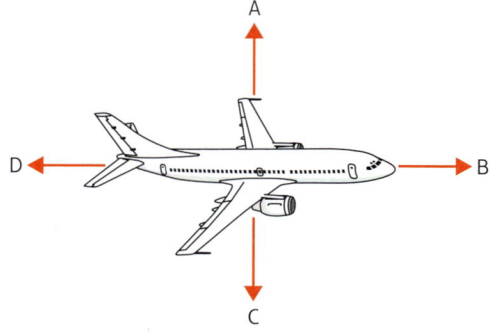

Also known as 'lift', this is the upward force that keeps the aircraft up in the air.

A upward thrust
B forward thrust
C weight
D air resistance

Also known as 'thrust', this is the force that propels the aircraft forward.

This is the gravitational force acting downward on the aircraft.

Also known as 'drag', this is the force that acts in the **opposite** direction to the forward thrust.

Exercise 1C

1 Convert to SI units:
   a 65 km h^{-1}
   b 15 g cm^{-2}
   c 30 cm per minute
   d 24 g m^{-3}
   e $4.5 \times 10^{-2} \text{ g cm}^{-3}$
   f $6.3 \times 10^{-3} \text{ kg cm}^{-2}$

2 Write down the names of the forces shown in each of these diagrams.
   a A box being pushed along rough ground
   b A dolphin swimming through the water

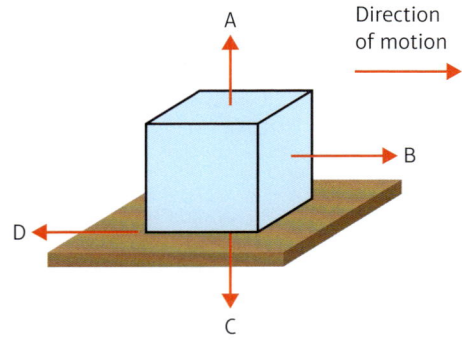

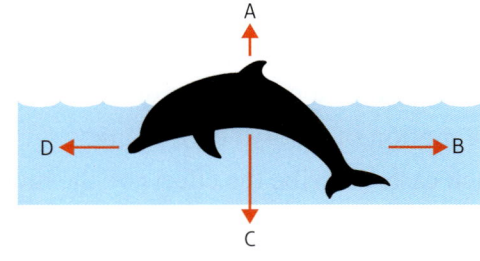

c A toy duck being pulled along by a string   d A man sliding down a hill on a sledge

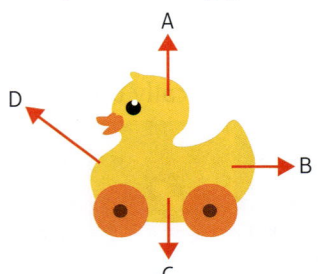
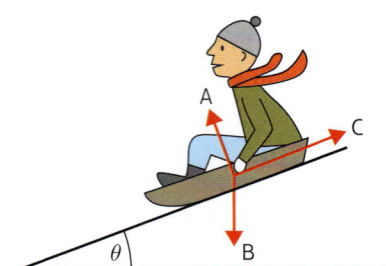

Chapter review 1

1 The motion of a cricket ball after it is hit until it lands on the cricket pitch can be modelled using the equation $h = \frac{1}{10}(24x - 3x^2)$, where h m is the vertical height of the ball above the cricket pitch and x m is the horizontal distance from where it was hit. Find:

Hint The path of the cricket ball is modelled as a quadratic curve. Draw a sketch for the model and use the symmetry of the curve.

  a the vertical height of the ball when it is at a horizontal distance of 2 m from where it was hit
  b the two horizontal distances for which the height of the ball was 2.1 m.

Given that the model is valid from when the ball is hit to when it lands on the cricket pitch:

  c find the values of x for which the model is valid
  d work out the **maximum** height of the cricket ball.

2 A diver dives from a diving board into a swimming pool with a depth of 4.5 m. The height of the diver above the water, h m, can be modelled using $h = 10 - 0.58x^2$ for $0 \leq x \leq 5$, where x m is the horizontal distance from the end of the diving board.

  a Find the height of the diver when $x = 2$ m.
  b Find the horizontal distance from the end of the diving board to the point where the diver enters the water.

In this model the diver is modelled as a particle.

  c Describe the effects of this modelling assumption.
  d Comment on the validity of this modelling assumption for the motion of the diver after she enters the water.

3 Make a list of the assumptions you might make to create simple models of the following:
  a the motion of a man skiing down a snow-covered slope
  b the motion of a yo-yo on a string.

In each case, describe the effects of the modelling assumptions.

4 Convert to SI units:

a 2.5 km per minute b 0.6 kg cm^{-2} c 1.2×10^3 g cm^{-3}

5 A man throws a bowling ball in a bowling alley.

a Make a list of the assumptions you might make to create a simple model of the motion of the bowling ball.

b Taking the direction in which the ball travels as the positive direction, state with a reason whether each of the following are likely to be positive or negative:

 i the velocity ii the acceleration.

Summary of key points

1 Mathematical models can be constructed to simulate real-life situations.

2 Modelling assumptions can be used to simplify your calculations.

3 The base SI units most commonly used in mechanics are:

Quantity	Unit	Symbol
Mass	kilogram	kg
Length/displacement	metre	m
Time	second	s

4 The derived SI units most commonly used in mechanics are:

Quantity	Unit	Symbol
Speed/velocity	metres per second	m s^{-1}
Acceleration	metres per second per second	m s^{-2}
Weight/force	newton	N (= kg m s^{-2})

2 CONSTANT ACCELERATION

3.1

Learning objectives

After completing this chapter you should be able to:

- Understand and interpret displacement–time graphs → pages 11–13
- Understand and interpret velocity–time graphs → pages 13–16
- Derive the constant acceleration formulae and use them to solve problems → pages 17–29
- Use the constant acceleration formulae to solve problems involving vertical motion under gravity → pages 29–35

Prior knowledge check

1. For each graph find:
   i the gradient
   ii the shaded area under the graph.

   a 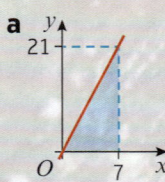　b 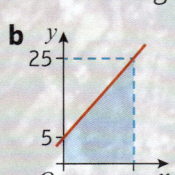　c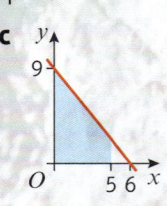

   ← International GCSE Mathematics

2. A car travels for 45 minutes at an average speed of 35 km h^{-1}. Find the distance travelled.
   ← International GCSE Mathematics

3. a Solve these simultaneous equations:
   $3x - 2y = 9$
   $x + 4y + 4 = 0$
   b Solve $2x^2 + 3x - 7 = 0$. Give your answers to 3 s.f.
   ← International GCSE Mathematics

A body falling freely under **gravity** can be modelled as having **constant acceleration**. You can use this to estimate the time that a cliff diver spends in freefall.

CONSTANT ACCELERATION   CHAPTER 2

2.1 Displacement–time graphs

You can represent the motion of an object on a **displacement–time graph**. Displacement is always plotted on (i.e. marked on) the vertical axis and time on the horizontal axis.

In these graphs, s represents the displacement of an object from a given point in metres and t represents the time taken in seconds.

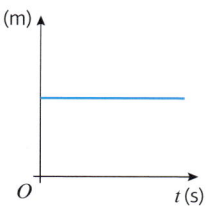

There is no change in the displacement over time and the object is stationary (not moving).

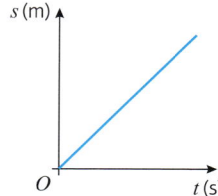

The displacement increases at a constant rate over time and the object is moving with constant velocity.

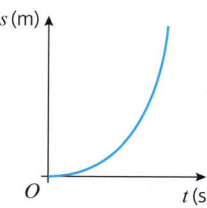

The displacement is increasing at a greater rate as time increases. The velocity is increasing and the object is accelerating.

- Velocity is the rate of change of displacement.
   - On a displacement–time graph the gradient represents the velocity.
   - If the displacement–time graph is a straight line, then the velocity is constant.

- Average velocity = $\dfrac{\text{displacement from starting point}}{\text{time taken}}$

- Average speed = $\dfrac{\text{total distance travelled}}{\text{time taken}}$

Example 1   SKILLS   INTERPRETATION

A cyclist rides in a straight line for 20 minutes. She waits for half an hour, then returns in a straight line to her starting point in 15 minutes. This is a displacement–time graph for her journey.

a Work out the average velocity for each stage of the journey in km h^{-1}.

b Write down the average velocity for the whole journey.

c Work out the average speed for the whole journey.

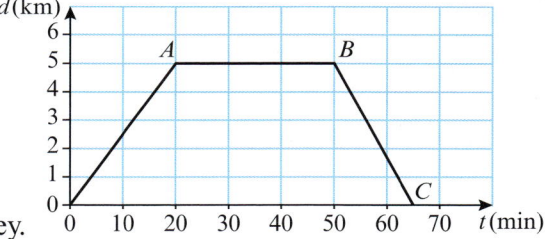

a Journey from O to A: time = 20 min; displacement = 5 km

Average velocity = $\dfrac{5}{20}$ = 0.25 km min^{-1}

To convert from km min^{-1} to km h^{-1} multiply by 60.

0.25 × 60 = 15 km h^{-1}

Journey from A to B: no change in displacement so average velocity = 0

A horizontal line on the graph indicates the cyclist is stationary.

Journey from B to C: time = 15 min; displacement = −5 km

Average velocity = $\dfrac{-5}{15}$ = $-\dfrac{1}{3}$ km min^{-1}

$-\dfrac{1}{3}$ × 60 = −20 km h^{-1}

The cyclist starts with a displacement of 5 km and finishes with a displacement of 0 km, so the change in displacement is −5 km, and velocity will be negative.

b The displacement for the whole journey is 0 so average velocity is 0.

c Total time = 65 min
Total distance travelled is 5 + 5 = 10 km
Average speed = $\frac{10}{65} = \frac{2}{13}$ km min^{-1}
$\frac{2}{13} \times 60 = 9.23$ km h^{-1} (3 s.f.)

At C the cyclist has returned to the starting point.

The cyclist has travelled 5 km away from the starting point and then 5 km back to the starting point.

Exercise 2A — SKILLS: INTERPRETATION

1 This is a displacement–time graph for a car travelling along a straight road. The journey is divided into 5 stages labelled A to E.
  a Work out the average velocity for each stage of the journey.
  b State the average velocity for the whole journey.
  c Work out the average speed for the whole journey.

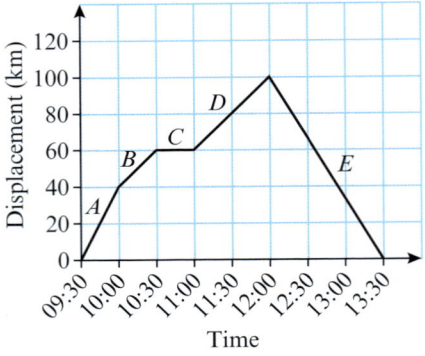

P 2 Khalid drives from his home to a hotel. He drives for $2\frac{1}{2}$ hours at an average velocity of 60 km h^{-1}. He then stops for lunch before continuing to his hotel. The diagram shows a displacement–time graph for Khalid's journey.
  a Work out the displacement of the hotel from Khalid's home.
  b Work out Khalid's average velocity for his whole journey.

Problem-solving
You need to work out the scale on the vertical axis.

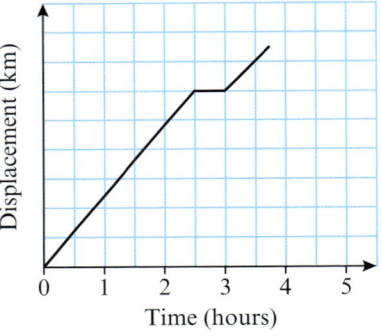

P 3 Sarah left home at 10:00 and cycled north in a straight line. The diagram shows a displacement–time graph for her journey.
  a Work out Sarah's velocity between 10:00 and 11:00.
  On her return journey, Sarah continued past her home for 3 km before returning.
  b Estimate the time that Sarah passed her home.
  c Work out Sarah's velocity for each of the last two stages of her journey.
  d Calculate Sarah's average speed for her entire journey.

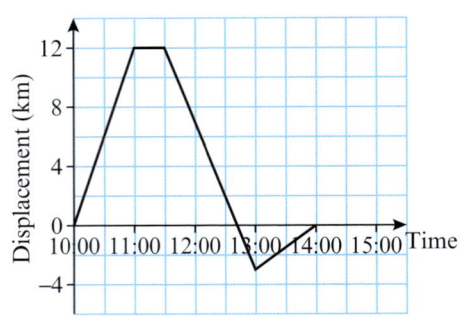

CONSTANT ACCELERATION — CHAPTER 2

P **4** A ball is thrown vertically up in the air and falls to the ground. This is a displacement–time graph for the motion of the ball.
  a Find the maximum height of the ball and the time at which it reaches that height.
  b Write down the velocity of the ball when it reaches its highest point.
  c Describe the motion of the ball:
   i from the time it is thrown to the time it reaches its highest point
   ii after reaching its highest point.

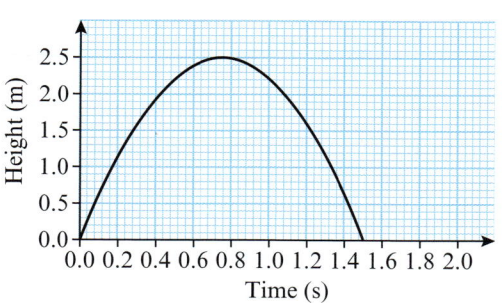

Hint To describe the motion you should state the direction of travel of the ball and whether it is accelerating or decelerating.

2.2 Velocity–time graphs

You can represent the motion of an object on a **velocity–time graph**. Velocity is always plotted on the vertical axis and time on the horizontal axis.

In these graphs v represents the velocity of an object in metres per second and t represents the time taken in seconds.

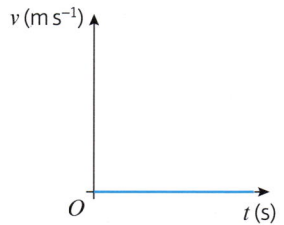 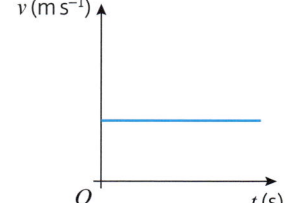 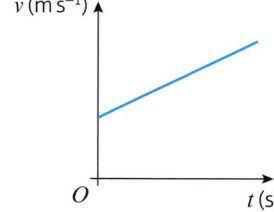

The velocity is zero and the object is stationary.

The velocity is unchanging and the object is moving with constant velocity.

The velocity is increasing at a constant rate and the object is moving with constant acceleration.

- **Acceleration is the rate of change of velocity.**
  - **In a velocity–time graph the gradient represents the acceleration.**
  - **If the velocity–time graph is a straight line, then the acceleration is constant.**

Notation Negative acceleration is sometimes described as deceleration or retardation.

This velocity–time graph represents the motion of an object travelling in a straight line at constant velocity V m s^{-1} for time T seconds.

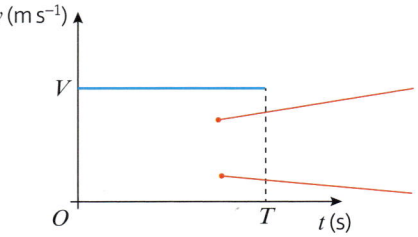

Area under the graph = $V \times T$

For an object with constant velocity, displacement = velocity × time

- **The area between a velocity–time graph and the horizontal axis represents the distance travelled.**
  - **For motion in a straight line with positive velocity, the area under the velocity–time graph up to a point t represents the displacement at time t.**

Example 2 — SKILLS: INTERPRETATION

The figure shows a velocity–time graph illustrating the motion of a cyclist moving along a straight road for a period of 12 seconds. For the first 8 seconds, she moves at a constant rate of $6\,\mathrm{m\,s^{-1}}$. She then decelerates at a constant rate, stopping after a further 4 seconds.

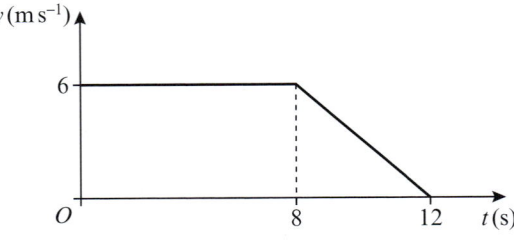

a Find the displacement from the starting point of the cyclist after this 12 second period.

b Work out the rate at which the cyclist decelerates.

a The displacement s after 12 s is given by the area under the graph.

$s = \tfrac{1}{2}(a + b)h$

$ = \tfrac{1}{2}(8 + 12) \times 6$

$ = 10 \times 6 = 60$

The displacement of the cyclist after 12 s is 60 m.

> Model the cyclist as a particle moving in a straight line.
>
> The displacement is represented by the area of the trapezium with these sides.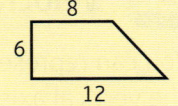
>
> You can use the formula for the area of a trapezium to calculate this area.

b The acceleration is the gradient of the slope.

$a = \dfrac{-6}{4} = -1.5$

The deceleration is $1.5\,\mathrm{m\,s^{-2}}$.

> The gradient is given by
> $\dfrac{\text{difference in the } v\text{-coordinates}}{\text{difference in the } t\text{-coordinates}}$
>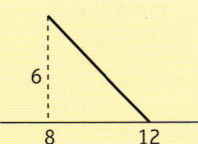

Example 3 — SKILLS: INTERPRETATION

A particle moves along a straight line. The particle accelerates **uniformly** from rest to a velocity of $8\,\mathrm{m\,s^{-1}}$ in T seconds. The particle then travels at a constant velocity of $8\,\mathrm{m\,s^{-1}}$ for $5T$ seconds. The particle then decelerates uniformly to rest in a further 40 seconds.

a Sketch a speed–time graph to illustrate the motion of the particle.

b Given that the total displacement of the particle is 600 m, find the value of T.

a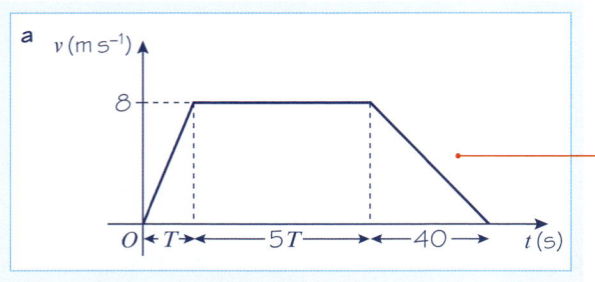

> If the particle accelerates from rest and decelerates to rest this means the initial and final velocities are zero.

Online   Explore how the area of the trapezium changes as the value of T changes using technology.

b The area of the trapezium is:

$s = \frac{1}{2}(a+b)h$

$= \frac{1}{2}(5T + 6T + 40) \times 8$

$= 4(11T + 40)$

The displacement is 600 m.

$4(11T + 40) = 600$

$44T + 160 = 600$

$T = \dfrac{600 - 160}{44} = 10\,s$

The length of the shorter of the two parallel sides is $5T$. The length of the longer side is $T + 5T + 40 = 6T + 40$.

Problem-solving

The displacement is equal to the area of the trapezium. Write an equation and solve it to find T.

Exercise 2B   SKILLS   INTERPRETATION, PROBLEM-SOLVING

1 The diagram shows the speed–time graph of the motion of an athlete running along a straight track. For the first 4 s, he accelerates uniformly from rest to a velocity of $9\,m\,s^{-1}$.
This velocity is then maintained for a further 8 s. Find:
   a the rate at which the athlete accelerates
   b the displacement from the starting point of the athlete after 12 s.

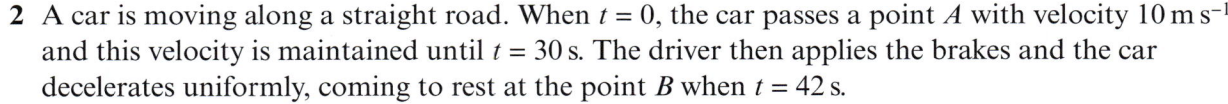

2 A car is moving along a straight road. When $t = 0$, the car passes a point A with velocity $10\,m\,s^{-1}$ and this velocity is maintained until $t = 30\,s$. The driver then applies the brakes and the car decelerates uniformly, coming to rest at the point B when $t = 42\,s$.
   a Sketch a velocity–time graph to illustrate the motion of the car.
   b Find the distance from A to B.

(E) 3 The diagram shows the velocity–time graph of the motion of a cyclist riding along a straight road. She accelerates uniformly from rest to $8\,m\,s^{-1}$ in 20 s. She then travels at a constant velocity of $8\,m\,s^{-1}$ for 40 s. She then decelerates uniformly to rest in 15 s. Find:

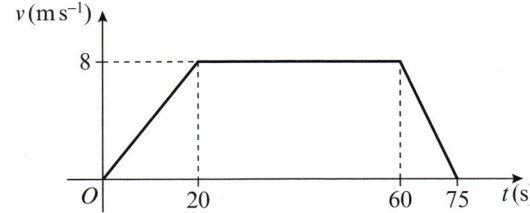

   a the acceleration of the cyclist in the first 20 s of motion   (2 marks)
   b the deceleration of the cyclist in the last 15 s of motion   (2 marks)
   c the displacement from the starting point of the cyclist after 75 s.   (2 marks)

(E) 4 A motorcyclist starts from rest at a point S on a straight race track. He moves with constant acceleration for 15 s, reaching a velocity of $30\,m\,s^{-1}$. He then travels at a constant velocity of $30\,m\,s^{-1}$ for T seconds. Finally he decelerates at a constant rate coming to rest at a point F, 25 s after he begins to decelerate.
   a Sketch a speed–time graph to illustrate the motion.   (3 marks)
   b Given that the distance between S and F is 2.4 km, calculate the time the motorcyclist takes to travel from S to F.   (3 marks)

5 A train starts from a station X and moves with constant acceleration of $0.6\,\text{m s}^{-2}$ for $20\,\text{s}$. The velocity it has reached after $20\,\text{s}$ is then maintained for T seconds. The train then decelerates from this velocity to rest in a further $40\,\text{s}$, stopping at a station Y.
  a Sketch a velocity–time graph to illustrate the motion of the train. **(3 marks)**
Given that the distance between the stations is $4.2\,\text{km}$, find:
  b the value of T **(3 marks)**
  c the distance travelled by the train while it is moving with constant velocity. **(2 marks)**

6 A particle moves along a straight line. The particle accelerates from rest to a velocity of $10\,\text{m s}^{-1}$ in $15\,\text{s}$. The particle then moves at a constant velocity of $10\,\text{m s}^{-1}$ for a period of time. The particle then decelerates uniformly to rest. The period of time for which the particle is travelling at a constant velocity is 4 times the period of time for which it is decelerating.
  a Sketch a speed–time graph to illustrate the motion of the particle. **(3 marks)**
  b Given that the displacement from the starting point of the particle after it comes to rest is $480\,\text{m}$, find the total time for which the particle is moving. **(3 marks)**

7 A particle moves $100\,\text{m}$ in a straight line. The diagram is a sketch of a velocity–time graph of the motion of the particle. The particle starts with velocity $u\,\text{m s}^{-1}$ and accelerates to a velocity of $10\,\text{m s}^{-1}$ in $3\,\text{s}$. The velocity of $10\,\text{m s}^{-1}$ is maintained for $7\,\text{s}$ and then the particle decelerates to rest in a further $2\,\text{s}$. Find:

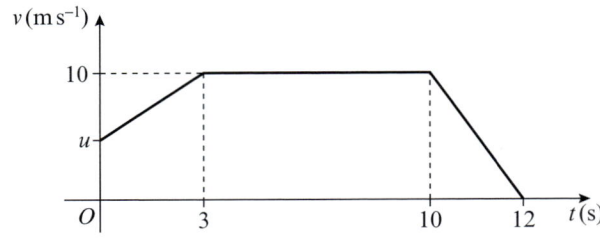

  a the value of u **(3 marks)**
  b the acceleration of the particle in the first $3\,\text{s}$ of motion. **(3 marks)**

8 A motorcyclist M leaves a road junction at time $t = 0$. She accelerates from rest at a rate of $3\,\text{m s}^{-2}$ for $8\,\text{s}$ and then maintains the velocity she has reached. A car C leaves the same road junction as M at time $t = 0$. The car accelerates from rest to $30\,\text{m s}^{-1}$ in $20\,\text{s}$ and then maintains the velocity of $30\,\text{m s}^{-1}$. C passes M as they both pass a pedestrian.
  a On the same diagram, sketch speed–time graphs to illustrate the motion of M and C. **(3 marks)**
  b Find the distance of the pedestrian from the road junction. **(3 marks)**

Challenge

The graph shows the velocity of an object travelling in a straight line during a 10-second time interval.
  a After how long did the object change direction?
  b Work out the total distance travelled by the object.
  c Work out the displacement from the starting point of the object after:
    i 6 seconds   **ii** 10 seconds.

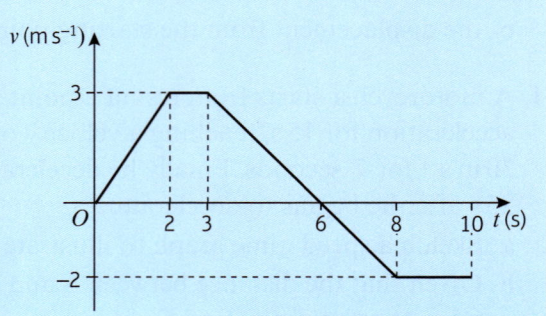

CONSTANT ACCELERATION   CHAPTER 2

2.3 Acceleration-time graphs

Acceleration is the rate of change of velocity.

You can represent the motion of an object on an acceleration-time graph. Acceleration is always plotted on the horizontal axis, and time on the vertical axis.

In the following graphs, a represents the acceleration of an object in m s^{-2} and t represents the time in seconds.

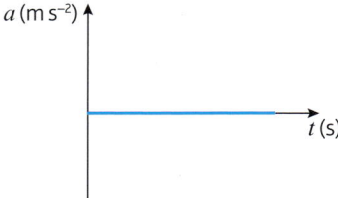

Zero acceleration – constant velocity

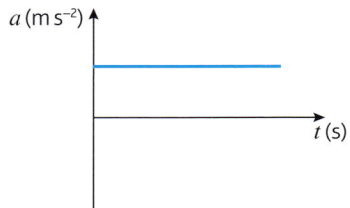

Positive acceleration; increase in velocity

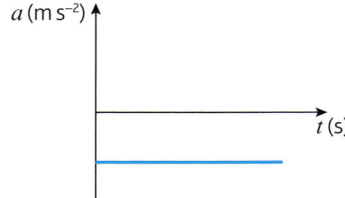

Negative acceleration (deceleration); decrease in velocity

A sloping line on an acceleration-time graph represents changing acceleration and is beyond the scope of this syllabus.

- The area between the acceleration-time graph and the horizontal axis represents the **change in velocity**. In other words, the area under the acceleration-time graph for a certain interval is equal to the change in velocity during that time interval.

Example 4   SKILLS   PROBLEM-SOLVING

The acceleration-time graph shows the motion of a car over a 20 second period.

a Calculate the change in velocity between $t = 0$ and $t = 10$ seconds.

b Given that the car was initially moving at 15 ms^{-1}, calculate the velocity after 16 seconds.

c Draw a velocity-time graph of the motion of the car.

d Find the distance travelled in the first 16 seconds.

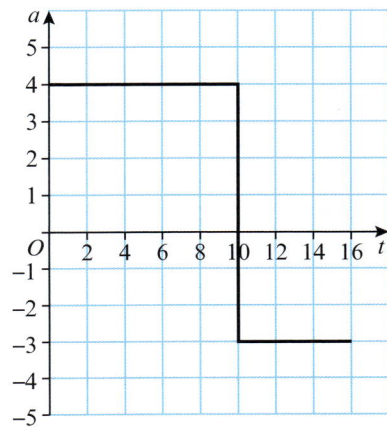

a Change in velocity = 4 × 10 = 40 ms⁻¹ — The change in velocity is the area under the line.

b Change in velocity from $t = 10$ to $t = 15$ is:
   Change = $-3 \times 6 = -18$ ms⁻¹ — Since the car is slowing down, this area is negative.
   Initial velocity $v_0 = 15$ ms⁻¹
   Velocity after 15 seconds
   $= 15 + 40 - 18 = 37$ ms⁻¹ — The final velocity is the sum of the changes and the initial velocity.

c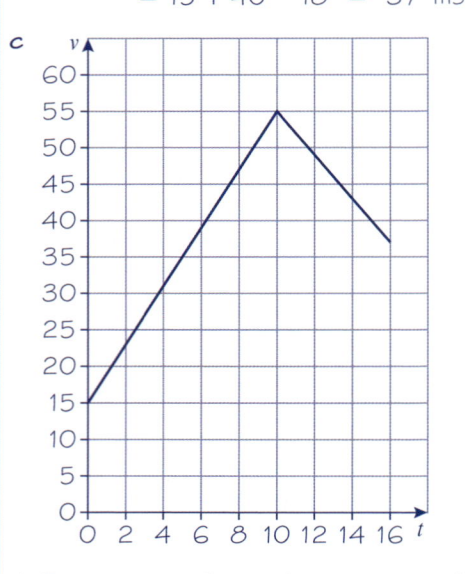

d Distance travelled is the area under the graph.
   Dist $= \frac{10}{2} \times (15 + 55) + \frac{6}{2} \times (15 + 37)$
   $= 506$ m

The distance travelled is the area under the graph of velocity. This area has been broken down into two trapeziums.

Exercise 2C SKILLS INTERPRETATION

1 The acceleration-time graph shows the motion of a racing car over 10 seconds of a race.
   Describe what is happening between:
   a $t = 0$ and $t = 2$ s
   b $t = 2$ and $t = 4$ s
   c $t = 4$ and $t = 8$ s.
   d Given that the initial velocity of the racing car was 44 m s⁻¹, its velocity at $t = 8$ s.

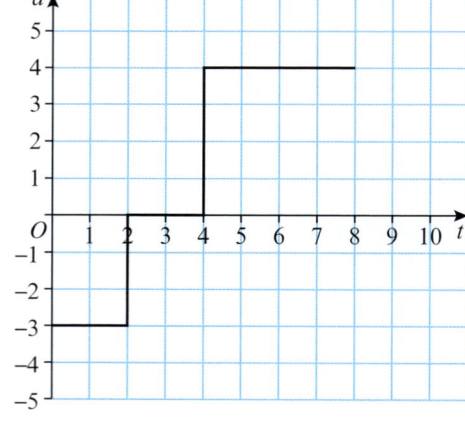

2 A motorcycle accelerates at a rate of 1 m s⁻² for 4 seconds, before travelling at a constant velocity for 6 seconds. It then decelerates at a rate of 0.5 m s⁻² for 4 seconds.
   a Draw an acceleration-time graph to show the motion of the car.
   Given that the initial velocity of the motorcycle was 18 m s⁻¹:
   b draw a velocity-time graph of the journey of the motorcycle from $t = 0$ to $t = 14$ seconds
   c hence, calculate the distance travelled by the motorcycle from $t = 0$ to $t = 14$ seconds.

3 The velocity-time graph shows the motion of a car over a 20 second period. Draw an acceleration-time graph to show the motion of the car.

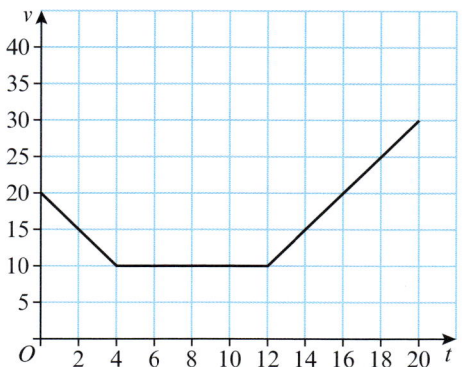

Challenge

A particle is travelling along the x-axis. At time $t = 0$, the velocity of the particle is 8 m s^{-1}, and the particle is accelerating at 5 m s^{-2}. After T seconds, the particle has constant velocity, and after $(T + 6)$ seconds the particle accelerates at -4 m s^{-2}.

Given that the velocity of the particle after $(T + 16)$ seconds is -12 m s^{-1}:

a draw an acceleration-time graph of the motion of the particle, giving your times in terms of T

b find the value of T.

Hint A negative acceleration means the particle is slowing down, and a negative velocity means the particle is travelling backwards.

2.4 Constant acceleration formulae 1

A standard set of letters is used for the motion of an object moving in a straight line with constant acceleration.

- s is the displacement.
- u is the initial velocity.
- v is the final velocity.
- a is the acceleration.
- t is the time.

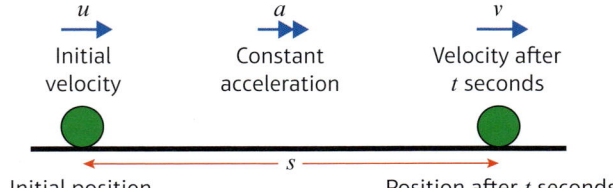

You can use these letters to label a velocity–time graph representing the motion of a particle moving in a straight line, accelerating from velocity u at time 0 to velocity v at time t.

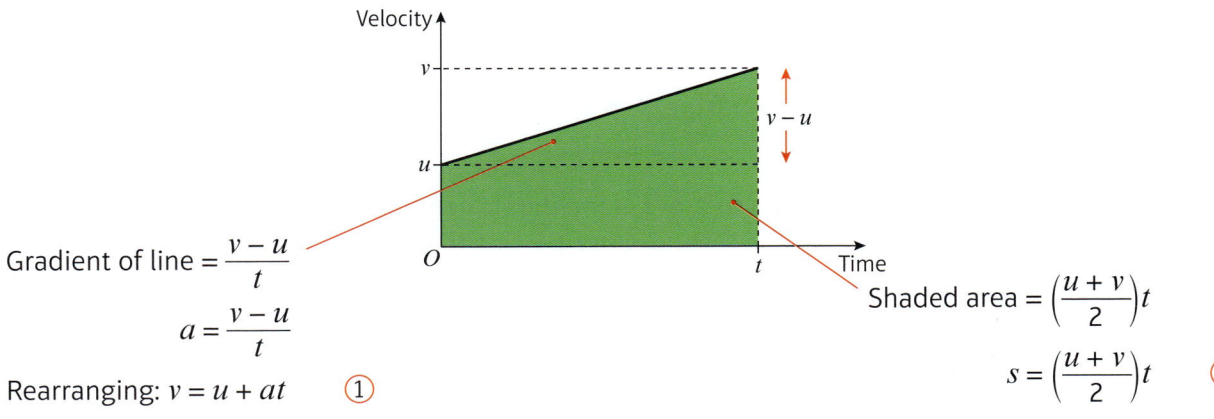

Gradient of line $= \dfrac{v - u}{t}$

$a = \dfrac{v - u}{t}$

Rearranging: $v = u + at$ ①

Shaded area $= \left(\dfrac{u + v}{2}\right)t$

$s = \left(\dfrac{u + v}{2}\right)t$ ②

- $v = u + at$ ①
- $s = \left(\dfrac{u + v}{2}\right)t$ ②

You need to know how to derive these formulae from the velocity–time graph.

Example 5 — SKILLS: PROBLEM-SOLVING

A cyclist is travelling along a straight road. She accelerates at a constant rate from a velocity of 4 m s⁻¹ to a velocity of 7.5 m s⁻¹ in 40 seconds. Find:

a the distance she travels in these 40 seconds

b her acceleration in these 40 seconds.

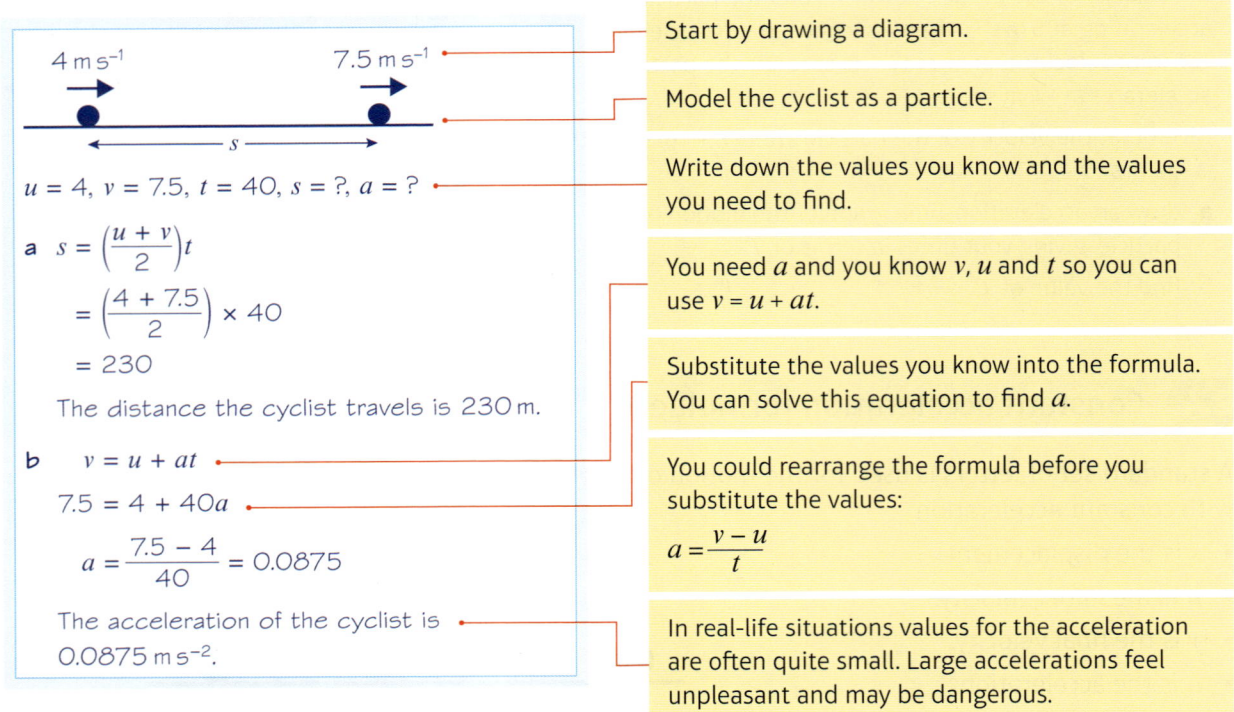

$u = 4$, $v = 7.5$, $t = 40$, $s = ?$, $a = ?$

a $s = \left(\dfrac{u+v}{2}\right)t$

$= \left(\dfrac{4 + 7.5}{2}\right) \times 40$

$= 230$

The distance the cyclist travels is 230 m.

b $v = u + at$

$7.5 = 4 + 40a$

$a = \dfrac{7.5 - 4}{40} = 0.0875$

The acceleration of the cyclist is 0.0875 m s^{-2}.

- Start by drawing a diagram.
- Model the cyclist as a particle.
- Write down the values you know and the values you need to find.
- You need a and you know v, u and t so you can use $v = u + at$.
- Substitute the values you know into the formula. You can solve this equation to find a.
- You could rearrange the formula before you substitute the values: $a = \dfrac{v - u}{t}$
- In real-life situations values for the acceleration are often quite small. Large accelerations feel unpleasant and may be dangerous.

Example 6 — SKILLS: PROBLEM-SOLVING

A particle moves in a straight line from a point A to a point B with constant deceleration 1.5 m s^{-2}. The velocity of the particle at A is 8 m s⁻¹ and the velocity of the particle at B is 2 m s⁻¹. Find:

a the time taken for the particle to move from A to B

b the distance from A to B.

After reaching B, the particle continues to move along the straight line with constant deceleration 1.5 m s^{-2}. The particle is at the point C 6 seconds after passing through the point A. Find:

c the velocity of the particle at C

d the distance from A to C.

CONSTANT ACCELERATION — CHAPTER 2

$u = 8$, $v = 2$, $a = -1.5$, $t = ?$, $s = ?$

a $v = u + at$
$2 = 8 - 1.5t$
$1.5t = 8 - 2$
$t = \dfrac{8 - 2}{1.5} = 4$

The time taken to move from A to B is 4 s.

b $s = \left(\dfrac{u + v}{2}\right)t$
$= \left(\dfrac{8 + 2}{2}\right) \times 4 = 20$

The distance from A to B is 20 m.

c $u = 8$, $a = -15$, $t = 6$, $v = ?$
$v = u + at$
$= 8 + (-1.5) \times 6$
$= 8 - 9 = -1$

The velocity of the particle is $1\,\text{m s}^{-1}$ in the direction $\overrightarrow{BA}$.

d $s = \left(\dfrac{u + v}{2}\right)t$
$= \left(\dfrac{8 + (-1)}{2}\right) \times 6$

The distance from A to C is 21 m.

Problem-solving

It's always a good idea to draw a sketch showing the positions of the particle. Mark the positive direction on your sketch, and remember that when the particle is **decelerating**, your value of a will be **negative**.

- Use your answer from part **a** as the value of t.
- The velocity at C is negative. This means that the particle is moving from right to left.
- Remember that to specify a velocity it is necessary to give speed and direction.
- Make sure you use the correct sign when substituting a negative value into a formula.

Example 7   SKILLS   PROBLEM-SOLVING

A car moves from traffic lights along a straight road with constant acceleration. The car starts from rest at the traffic lights and 30 seconds later the car passes a speed-trap where it is registered as travelling at 45 km h⁻¹. Find:

a the acceleration of the car

b the distance between the traffic lights and the speed-trap.

Hint Convert all your measurements into base SI units before substituting values into the formulae.

$45\,\text{km h}^{-1} = 45 \times \dfrac{1000}{3600}\,\text{m s}^{-1} = 12.5\,\text{m s}^{-1}$

$u = 0$, $v = 12.5$, $t = 30$, $a = ?$, $s = ?$

- Convert into SI units, using:
  1 km = 1000 m
  1 hour = 60 × 60 s = 3600 s
- Model the car as a particle and draw a diagram.
- The car starts from rest, so the initial velocity is zero.

> **a** $v = u + at$
> $12.5 = 0 + 30a$
> $a = \frac{12.5}{30} = \frac{5}{12}$
> The acceleration of the car is $\frac{5}{12}$ m s^{-2}. ← This is an exact answer. If you want to give an answer using decimals, you should round to three significant figures.
>
> **b** $s = \left(\frac{u+v}{2}\right)t$
> $= \left(\frac{0 + 12.5}{2}\right) \times 30 = 187.5$
> The distance between the traffic lights and the speed-trap is 187.5 m.

Exercise 2D — SKILLS — PROBLEM-SOLVING

1 A particle is moving in a straight line with constant acceleration $3\,\text{m s}^{-2}$.
At time $t = 0$, the speed of the particle is $2\,\text{m s}^{-1}$.
Find the speed of the particle at time $t = 6\,\text{s}$.

2 A car is approaching traffic lights. The car is travelling with speed $10\,\text{m s}^{-1}$. The driver applies the brakes to the car and the car comes to rest with constant deceleration in $16\,\text{s}$. Modelling the car as a particle, find the deceleration of the car.

3 A car accelerates uniformly while travelling on a straight road. The car passes two signposts $360\,\text{m}$ apart. The car takes $15\,\text{s}$ to travel from one signpost to the other. When passing the second signpost, it has speed $28\,\text{m s}^{-1}$. Find the speed of the car at the first signpost.

4 A cyclist is moving along a straight road from A to B with constant acceleration $0.5\,\text{m s}^{-2}$. Her speed at A is $3\,\text{m s}^{-1}$ and it takes her 12 seconds to cycle from A to B. Find:
  a her speed at B
  b the distance from A to B.

5 A particle is moving along a straight line with constant acceleration from a point A to a point B, where $AB = 24\,\text{m}$. The particle takes $6\,\text{s}$ to move from A to B and the speed of the particle at B is $5\,\text{m s}^{-1}$. Find:
  a the speed of the particle at A
  b the acceleration of the particle.

6 A particle moves in a straight line from a point A to a point B with constant deceleration $1.2\,\text{m s}^{-2}$. The particle takes $6\,\text{s}$ to move from A to B. The speed of the particle at B is $2\,\text{m s}^{-1}$ and the direction of motion of the particle has not changed. Find:
  a the speed of the particle at A
  b the distance from A to B.

CONSTANT ACCELERATION   CHAPTER 2

(P) **7** A train, travelling on a straight track, is slowing down with constant deceleration $0.6\,\text{m\,s}^{-2}$. The train passes one signal with speed $72\,\text{km\,h}^{-1}$ and a second signal 25 s later. Find:

Hint Convert the speeds into $\text{m\,s}^{-1}$ before substituting.

  a the speed, in $\text{km\,h}^{-1}$, of the train as it passes the second signal
  b the distance between the signals.

8 A particle moves in a straight line from a point A to a point B with a constant deceleration of $4\,\text{m\,s}^{-2}$. At A the particle has speed $32\,\text{m\,s}^{-1}$ and the particle comes to rest at B. Find:
  a the time taken for the particle to travel from A to B
  b the distance between A and B.

(E) **9** A skier travelling in a straight line up a hill experiences a constant deceleration. At the bottom of the hill, the skier has a speed of $16\,\text{m\,s}^{-1}$ and, after moving up the hill for 40 s, he comes to rest. Find:
  a the deceleration of the skier **(2 marks)**
  b the distance from the bottom of the hill to the point where the skier comes to rest. **(4 marks)**

(E) **10** A particle is moving in a straight line with constant acceleration. The points A, B and C lie on this line. The particle moves from A through B to C. The speed of the particle at A is $2\,\text{m\,s}^{-1}$ and the speed of the particle at B is $7\,\text{m\,s}^{-1}$. The particle takes 20 s to move from A to B.
  a Find the acceleration of the particle. **(2 marks)**
  The speed of the particle at C is $11\,\text{m\,s}^{-1}$. Find:
  b the time taken for the particle to move from B to C **(2 marks)**
  c the distance between A and C. **(3 marks)**

(E) **11** A particle moves in a straight line from A to B with constant acceleration $1.5\,\text{m\,s}^{-2}$. It then moves along the same straight line from B to C with a different acceleration. The speed of the particle at A is $1\,\text{m\,s}^{-1}$ and the speed of the particle at C is $43\,\text{m\,s}^{-1}$. The particle takes 12 s to move from A to B and 10 s to move from B to C. Find:
  a the speed of the particle at B **(2 marks)**
  b the acceleration of the particle as it moves from B to C **(2 marks)**
  c the distance from A to C. **(3 marks)**

(E/P) **12** A cyclist travels with constant acceleration $x\,\text{m\,s}^{-2}$, in a straight line, from rest to $5\,\text{m\,s}^{-1}$ in 20 s. She then decelerates from $5\,\text{m\,s}^{-1}$ to rest with constant deceleration $\frac{1}{2}x\,\text{m\,s}^{-2}$. Find:
  a the value of x **(2 marks)**
  b the total distance she travelled. **(4 marks)**

Problem-solving
You could sketch a velocity–time graph of the cyclist's motion and use the area under the graph to find the total distance travelled.

13 A particle is moving with constant acceleration in a straight line. It passes through three points, A, B and C, with speeds $20\,\text{m s}^{-1}$, $30\,\text{m s}^{-1}$ and $45\,\text{m s}^{-1}$ respectively. The time taken to move from A to B is t_1 seconds and the time taken to move from B to C is t_2 seconds.

a Show that $\dfrac{t_1}{t_2} = \dfrac{2}{3}$. **(3 marks)**

Given also that the total time taken for the particle to move from A to C is $50\,\text{s}$:

b find the distance between A and B. **(5 marks)**

Challenge

A particle moves in a straight line from A to B with constant acceleration. The particle moves from A with speed $3\,\text{m s}^{-1}$. It reaches point B with speed $5\,\text{m s}^{-1}$ t seconds later.

One second after the first particle leaves point A, a second particle also starts to move in a straight line from A to B with constant acceleration. Its speed at point A is $4\,\text{m s}^{-1}$ and it reaches point B with speed $8\,\text{m s}^{-1}$ at the same time as the first particle.

Find:

a the value of t

b the distance between A and B.

Problem-solving
The time taken for the second particle to travel from A to B is $(t - 1)$ seconds.

2.5 Constant acceleration formulae 2

You can use the formulae $v = u + at$ and $s = \left(\dfrac{u + v}{2}\right)t$ to work out three more formulae.

You can eliminate t from the formulae for constant acceleration.

$t = \dfrac{v - u}{a}$ ———— Rearrange the formula $v = u + at$ to make t the subject.

$s = \left(\dfrac{u + v}{2}\right)\left(\dfrac{v - u}{a}\right)$ ———— Substitute this expression for t into $s = \left(\dfrac{u + v}{2}\right)t$.

$2as = v^2 - u^2$

- $v^2 = u^2 + 2as$ ———— Multiply out the brackets and rearrange.

You can also eliminate v from the formulae for constant acceleration.

$s = \left(\dfrac{u + u + at}{2}\right)t$ ———— Substitute $v = u + at$ into $s = \left(\dfrac{u + v}{2}\right)t$.

$s = \left(\dfrac{2u}{2} + \dfrac{at}{2}\right)t$

$s = \left(u + \tfrac{1}{2}at\right)t$ ———— Multiply out the brackets and rearrange.

- $s = ut + \tfrac{1}{2}at^2$

Finally, you can eliminate u by substituting into this formula:

$$s = (v - at)t + \tfrac{1}{2}at^2$$

Substitute $u = v - at$ into $s = ut + \tfrac{1}{2}at^2$.

- $s = vt - \tfrac{1}{2}at^2$

■ You need to be able to recall and use the five formulae for solving problems about particles moving in a straight line with constant acceleration.

- $v = u + at$

- $s = \left(\dfrac{u+v}{2}\right)t$

- $v^2 = u^2 + 2as$

- $s = ut + \tfrac{1}{2}at^2$

- $s = vt - \tfrac{1}{2}at^2$

Watch out These five formulae are sometimes referred to as the **kinematics** formulae or **suvat** **formulae**.

Example 8 — SKILLS — PROBLEM-SOLVING

A particle is moving along a straight line from A to B with constant acceleration $5\,\text{m s}^{-2}$. The velocity of the particle at A is $3\,\text{m s}^{-1}$ in the direction $\overrightarrow{AB}$. The velocity of the particle at B is $18\,\text{m s}^{-1}$ in the same direction. Find the distance from A to B.

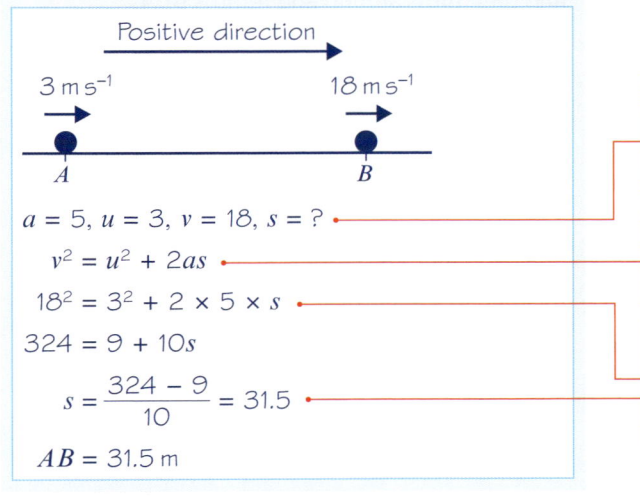

$a = 5,\ u = 3,\ v = 18,\ s = ?$

$v^2 = u^2 + 2as$

$18^2 = 3^2 + 2 \times 5 \times s$

$324 = 9 + 10s$

$s = \dfrac{324 - 9}{10} = 31.5$

$AB = 31.5\,\text{m}$

Write down the values you know and the values you need to find. This will help you choose the correct formula.

t is not involved so choose the formula that does not have t in it.

Substitute in the values you are given and solve the equation for s. This gives the distance you were asked to find.

Example 9 — SKILLS: PROBLEM-SOLVING

A particle is moving in a straight horizontal line with constant deceleration $4\,\text{m s}^{-2}$. At time $t = 0$ the particle passes through a point O with velocity $13\,\text{m s}^{-1}$ travelling toward a point A, where $OA = 20\,\text{m}$. Find:

a the times when the particle passes through A

b the value of t when the particle returns to O.

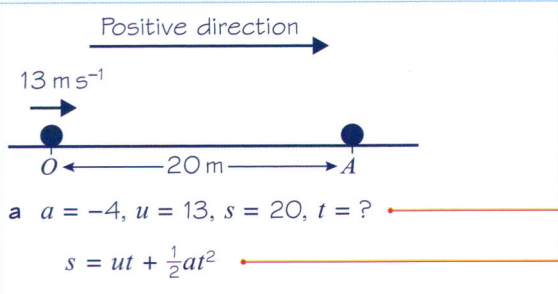

a $a = -4, u = 13, s = 20, t = ?$

$s = ut + \tfrac{1}{2}at^2$

$20 = 13t + \tfrac{1}{2} \times (-4)t^2$

$ = 13t - 2t^2$

$2t^2 - 13t + 20 = 0$

$(2t - 5)(t - 4) = 0$

$t = \tfrac{5}{2}$, or $t = 4$

The particle moves through A twice, $2\tfrac{1}{2}$ seconds and 4 seconds after moving through O.

b The particle returns to O when $s = 0$.

$s = 0, u = 13, a = -4, t = ?$

$s = ut + \tfrac{1}{2}at^2$

$0 = 13t - 2t^2$

$ = t(13 - 2t)$

$t = 0$, or $t = \tfrac{13}{2}$

The particle returns to O 6.5 seconds after it first passed through O.

The particle is decelerating so the value of a is negative.

You are told the values of a, u and s and asked to find t. You are given no information about v and are not asked to find it so you choose the formula without v.

Problem-solving

When you use $s = ut + \tfrac{1}{2}at^2$ with an unknown value of t you obtain a quadratic equation in t. You can solve this equation by factorising, or using the quadratic formula, $x = \dfrac{-b \pm \sqrt{b^2 - 4ac}}{2a}$.

There are two answers. Both are correct. The particle moves from O to A, goes beyond A and then turns around and returns to A.

When the particle returns to O, its displacement (distance) from O is zero.

The first solution ($t = 0$) represents the starting position of the particle. The other solution ($t = \tfrac{13}{2}$) tells you when the particle returns to O.

Example 10 — SKILLS: PROBLEM-SOLVING

A particle P is moving on the x-axis with constant deceleration $2.5\,\text{m s}^{-2}$. At time $t = 0$, the particle P passes through the origin O, moving in the positive direction of x with velocity $15\,\text{m s}^{-1}$. Find:

a the time between the instant when P first passes through O and the instant when it returns to O

b the total distance travelled by P during this time.

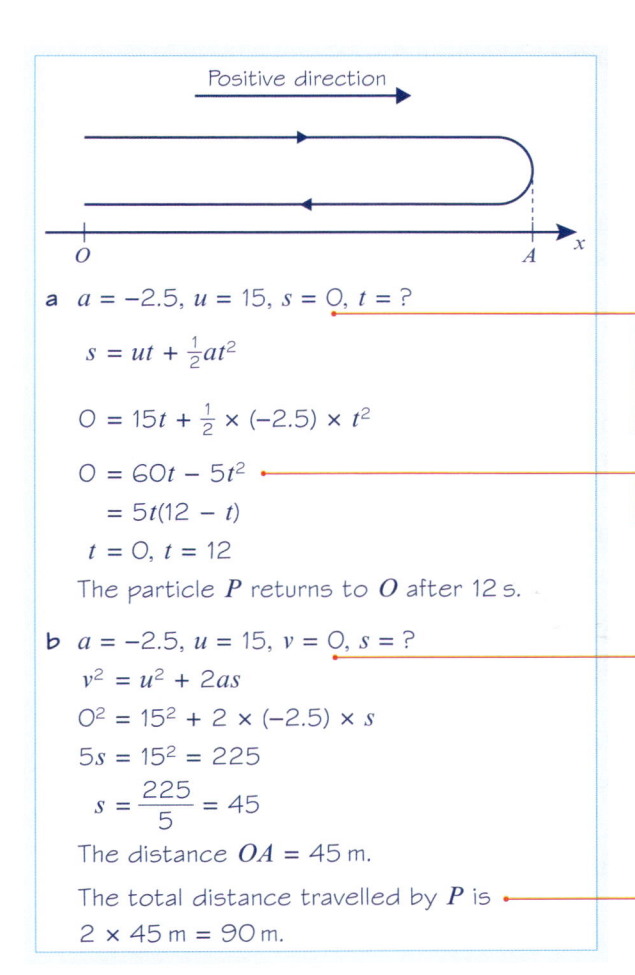

CONSTANT ACCELERATION — CHAPTER 2

Positive direction →

O ———————————————→ A x

a $a = -2.5$, $u = 15$, $s = 0$, $t = ?$

$s = ut + \frac{1}{2}at^2$

$0 = 15t + \frac{1}{2} \times (-2.5) \times t^2$

$0 = 60t - 5t^2$

$= 5t(12 - t)$

$t = 0$, $t = 12$

The particle P returns to O after 12 s.

b $a = -2.5$, $u = 15$, $v = 0$, $s = ?$

$v^2 = u^2 + 2as$

$0^2 = 15^2 + 2 \times (-2.5) \times s$

$5s = 15^2 = 225$

$s = \dfrac{225}{5} = 45$

The distance $OA = 45$ m.

The total distance travelled by P is

2×45 m $= 90$ m.

Problem-solving

Before you start, draw a sketch so you can see what is happening. The particle moves through O with a positive velocity. As it is decelerating it slows down and will eventually have zero velocity at a point A, which you don't yet know. As the particle is still decelerating, its velocity becomes negative, so the particle changes direction and returns to O.

When the particle returns to O, its displacement (distance) from O is zero.

Multiply by 4 to get whole-number coefficients.

At the furthest point from O, labelled A in the diagram, the particle changes direction. At that point, for an instant, the particle has zero velocity.

In the 12 s the particle has been moving it has travelled to A and back. The total distance travelled is twice the distance OA.

Exercise 2E — SKILLS — PROBLEM-SOLVING

1 A particle is moving in a straight line with constant acceleration 2.5 m s^{-2}. It passes a point A with velocity 3 m s^{-1} and later passes through a point B, where $AB = 8$ m. Find the velocity of the particle as it passes through B.

2 A car is accelerating at a constant rate along a straight horizontal road. Travelling at 8 m s^{-1}, it passes a lamp post and 6 s later it passes a sign. The distance between the lamp post and the sign is 60 m. Find the acceleration of the car.

3 A cyclist travelling at 12 m s^{-1} applies her brakes and comes to rest after travelling 36 m in a straight line. Assuming that the brakes cause the cyclist to decelerate uniformly, find the deceleration.

4 A train is moving along a straight horizontal track with constant acceleration. The train passes a signal with a velocity of 54 km h^{-1} and a second signal with a velocity of 72 km h^{-1}. The distance between the two signals is 500 m. Find, in m s^{-2}, the acceleration of the train.

5 A particle moves along a straight line, with constant acceleration, from a point A to a point B where $AB = 48$ m. At A the particle has velocity 4 m s^{-1} and at B it has velocity 16 m s^{-1}. Find:
   a the acceleration of the particle
   b the time the particle takes to move from A to B.

6 A particle moves along a straight line with constant acceleration $3\,\text{m}\,\text{s}^{-2}$. The particle moves 38 m in 4 s. Find:
  a the initial speed of the particle
  b the final speed of the particle.

7 The driver of a car is travelling at $18\,\text{m}\,\text{s}^{-1}$ along a straight road when she sees an obstruction ahead. She applies the brakes and the brakes cause the car to slow down to rest with a constant deceleration of $3\,\text{m}\,\text{s}^{-2}$. Find:
  a the distance travelled as the car decelerates
  b the time it takes for the car to decelerate from $18\,\text{m}\,\text{s}^{-1}$ to rest.

8 A stone is sliding across a frozen lake in a straight line. The initial speed of the stone is $12\,\text{m}\,\text{s}^{-1}$. The friction between the stone and the ice causes the stone to slow down at a constant rate of $0.8\,\text{m}\,\text{s}^{-2}$. Find:
  a the distance travelled by the stone before coming to rest
  b the speed of the stone at the instant when it has travelled half of this distance.

9 A particle is moving along a straight line OA with constant acceleration $2.5\,\text{m}\,\text{s}^{-2}$. At time $t = 0$, the particle passes through O with speed $8\,\text{m}\,\text{s}^{-1}$ and is moving in the direction OA. The distance OA is 40 m. Find:
  a the time taken for the particle to move from O to A
  b the speed of the particle at A. Give your answers to one decimal place.

10 A particle travels with uniform deceleration $2\,\text{m}\,\text{s}^{-2}$ in a horizontal line. The points A and B lie on the line and $AB = 32$ m. At time $t = 0$, the particle passes through A with speed $12\,\text{m}\,\text{s}^{-1}$ in the direction $\overrightarrow{AB}$. Find:
  a the values of t when the particle is at B
  b the speed of the particle for each of these values of t.

(E/P) 11 A particle is moving along the x-axis with constant deceleration $5\,\text{m}\,\text{s}^{-2}$. At time $t = 0$, the particle passes through the origin O with velocity $12\,\text{m}\,\text{s}^{-1}$ in the positive direction. At time t seconds, the particle passes through the point A with x-coordinate 8. Find:

> **Problem-solving**
>
> The particle will pass through A twice. Use $s = ut + \tfrac{1}{2}at^2$ to set up and solve a quadratic equation.

  a the values of t (3 marks)
  b the speed of the particle as it passes through the point with x-coordinate -8. (3 marks)

(E) 12 A particle P is moving on the x-axis with constant deceleration $4\,\text{m}\,\text{s}^{-2}$. At time $t = 0$, P passes through the origin O with speed $14\,\text{m}\,\text{s}^{-1}$ in the positive direction. The point A lies on the axis and $OA = 22.5$ m. Find:
  a the difference between the times when P passes through A (4 marks)
  b the total distance travelled by P during the interval between these times. (3 marks)

(E/P) 13 A car is travelling along a straight horizontal road with constant acceleration. The car passes over three consecutive points A, B and C where $AB = 100$ m and $BC = 300$ m. The speed of the car at B is $14\,\text{m}\,\text{s}^{-1}$ and the speed of the car at C is $20\,\text{m}\,\text{s}^{-1}$. Find:
  a the acceleration of the car (3 marks)
  b the time taken for the car to travel from A to C. (3 marks)

E/P 14 Two particles P and Q are moving along the same straight horizontal line with constant accelerations $2\,\text{m s}^{-2}$ and $3.6\,\text{m s}^{-2}$ respectively. At time $t = 0$, P passes through a point A with speed $4\,\text{m s}^{-1}$. One second later Q passes through A with speed $3\,\text{m s}^{-1}$, moving in the same direction as P.
   a Write down expressions for the displacements of P and Q from A, in terms of t, where t seconds is the time after P has passed through A. **(2 marks)**
   b Find the value of t where the particles meet. **(3 marks)**
   c Find the distance of A from the point where the particles meet. **(3 marks)**

> **Problem-solving**
> When P and Q meet, their displacements from A are equal.

E/P 15 In an orienteering competition, a competitor moves in a straight line past three checkpoints, P, Q and R, where $PQ = 2.4\,\text{km}$ and $QR = 11.5\,\text{km}$. The competitor is modelled as a particle moving with constant acceleration. She takes 1 hour to travel from P to Q and 1.5 hours to travel from Q to R. Find:
   a the acceleration of the competitor
   b her speed at the instant she passes P. **(7 marks)**

2.6 Vertical motion under gravity

You can use the formulae for constant acceleration to model an object moving vertically under gravity.

■ **The force of gravity causes all objects to accelerate toward the Earth. If you ignore the effects of air resistance, this acceleration is constant. It does not depend on the mass of the object.**

As the force of gravity does not depend on mass, this means that in a vacuum an apple and a feather would both accelerate downward at the same rate.

On Earth, the acceleration due to gravity is represented by the letter g and is approximately $9.8\,\text{m s}^{-2}$.

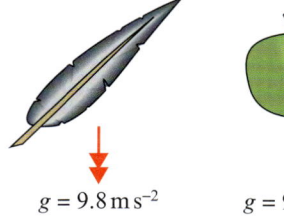

$g = 9.8\,\text{m s}^{-2}$ $g = 9.8\,\text{m s}^{-2}$

The actual value of the acceleration can vary by very small amounts in different places due to the changing radius of the earth and height above sea level.

■ **An object moving vertically under gravity can be modelled as a particle with a constant downward acceleration of $g = 9.8\,\text{m s}^{-2}$.**

> **Watch out** In mechanics questions you will always use $g = 9.8\,\text{m s}^{-2}$ unless a question specifies otherwise. However, if a different value of g is specified (e.g. $g = 10\,\text{m s}^{-2}$ or $g = 9.81\,\text{m s}^{-2}$) the degree of accuracy in your answer should be chosen to be consistent with this value.

When solving problems about vertical motion you can choose the positive direction to be either upward or downward. Acceleration due to gravity is always downward, so if the positive direction is upward then $g = -9.8\,\text{m s}^{-2}$.

> **Notation** The total time that an object is in motion from the time it is projected (thrown) upward to the time it hits the ground is called the **time of flight**. The initial speed is sometimes called the **speed of projection**.

Example 11 — SKILLS: PROBLEM-SOLVING

A book falls off the top shelf of a bookcase. The shelf is 1.4 m above a wooden floor. Find:

a the time the book takes to reach the floor

b the speed with which the book strikes the floor.

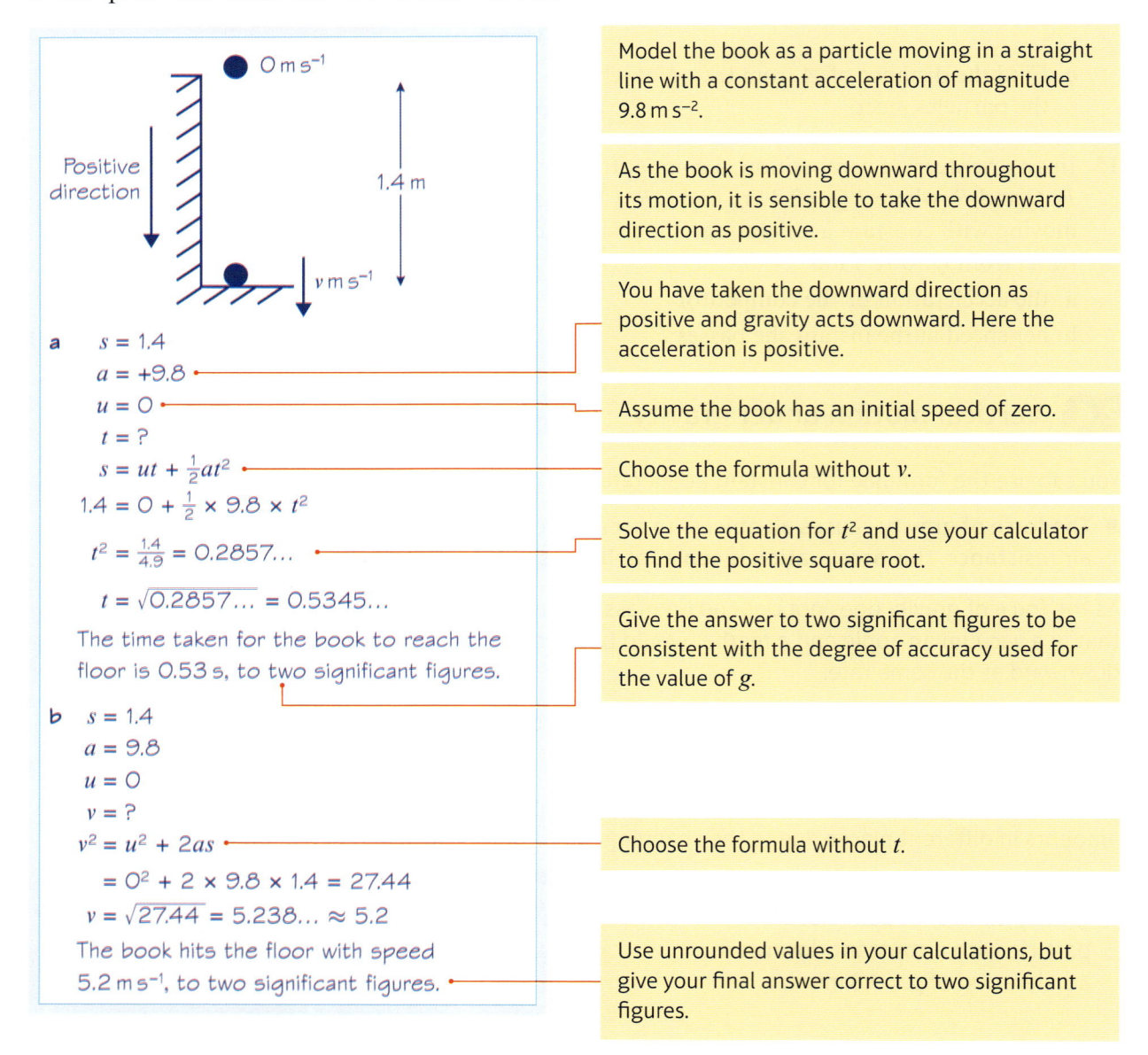

Model the book as a particle moving in a straight line with a constant acceleration of magnitude $9.8\,\text{m s}^{-2}$.

As the book is moving downward throughout its motion, it is sensible to take the downward direction as positive.

a $s = 1.4$
   $a = +9.8$ — *You have taken the downward direction as positive and gravity acts downward. Here the acceleration is positive.*
   $u = 0$ — *Assume the book has an initial speed of zero.*
   $t = ?$
   $s = ut + \tfrac{1}{2}at^2$ — *Choose the formula without v.*
   $1.4 = 0 + \tfrac{1}{2} \times 9.8 \times t^2$
   $t^2 = \tfrac{1.4}{4.9} = 0.2857\ldots$ — *Solve the equation for t^2 and use your calculator to find the positive square root.*
   $t = \sqrt{0.2857\ldots} = 0.5345\ldots$

The time taken for the book to reach the floor is 0.53 s, to two significant figures. — *Give the answer to two significant figures to be consistent with the degree of accuracy used for the value of g.*

b $s = 1.4$
   $a = 9.8$
   $u = 0$
   $v = ?$
   $v^2 = u^2 + 2as$ — *Choose the formula without t.*
   $= 0^2 + 2 \times 9.8 \times 1.4 = 27.44$
   $v = \sqrt{27.44} = 5.238\ldots \approx 5.2$

The book hits the floor with speed $5.2\,\text{m s}^{-1}$, to two significant figures. — *Use unrounded values in your calculations, but give your final answer correct to two significant figures.*

CONSTANT ACCELERATION — CHAPTER 2

Example 12 — SKILLS: PROBLEM-SOLVING

A ball is projected vertically upward, from a point X which is 7 m above the ground, with speed $21\,\text{m s}^{-1}$. Find:

a the greatest height above the ground reached by the ball

b the time of flight of the ball.

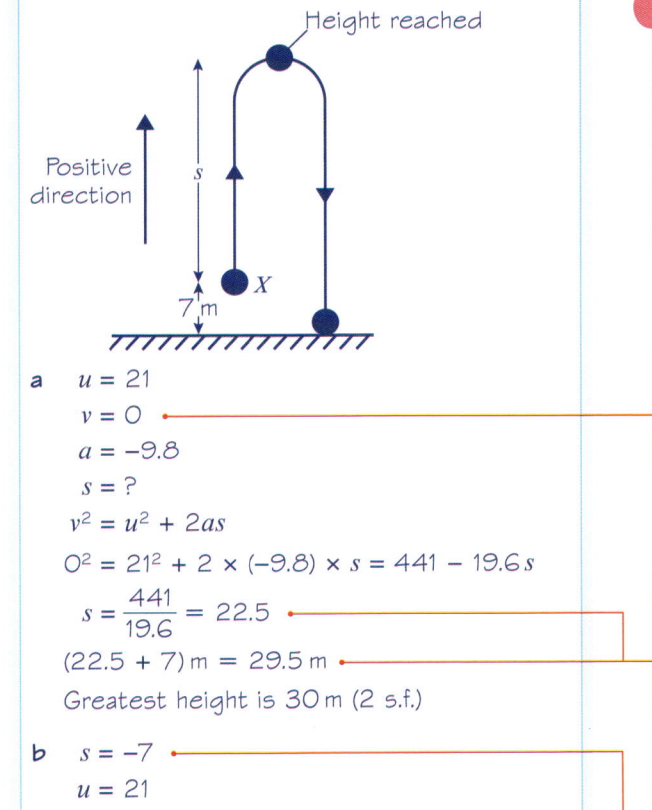

Problem-solving

In this sketch the upward and downward motions have been sketched side by side. In reality they would be on top of one another, but drawing them separately makes it easier to see what is going on. Remember that X is 7 m above the ground, so mark this height on your sketch.

a $u = 21$
$v = 0$
$a = -9.8$
$s = ?$
$v^2 = u^2 + 2as$
$0^2 = 21^2 + 2 \times (-9.8) \times s = 441 - 19.6s$
$s = \dfrac{441}{19.6} = 22.5$
$(22.5 + 7)\,\text{m} = 29.5\,\text{m}$
Greatest height is 30 m (2 s.f.)

At its highest point, the ball is turning around. For an instant, it is neither going up nor down, so its velocity is zero.

22.5 m is the distance the ball has moved above X, but X is 7 m above the ground. You must add on another 7 m to get the greatest height above the ground reached by the ball.

b $s = -7$
$u = 21$
$a = -9.8$
$t = ?$
$s = ut + \tfrac{1}{2}at^2$
$-7 = 21t - 4.9t^2$
$4.9t^2 - 21t - 7 = 0$
$t = \dfrac{-b \pm \sqrt{(b^2 - 4ac)}}{2a}$
$= \dfrac{-(-21) \pm \sqrt{((-21)^2 - 4 \times 4.9 \times (-7))}}{2 \times 4.9}$
$= \dfrac{21 \pm \sqrt{578.2}}{9.8} \approx \dfrac{21 \pm 24.046}{9.8}$
$t \approx 4.5965,$
or $t \approx -0.3108$
Time of flight is 4.6 s (2 s.f.)

The time of flight is the total time that the ball is in motion from the time that it is projected to the time that it stops moving. Here the ball will stop when it hits the ground. The point where the ball hits the ground is 7 m below the point from which it was projected so $s = -7$.

Online Use your calculator to check solutions to quadratic equations quickly.

Rearrange the equation and use the quadratic formula.

Take the positive answer and round to two significant figures.

Example 13 — SKILLS: PROBLEM-SOLVING

A particle is projected vertically upward from a point O with speed $u\,\text{m s}^{-1}$. The greatest height reached by the particle is 62.5 m above O. Find:

a the value of u

b the total time for which the particle is 50 m or more above O.

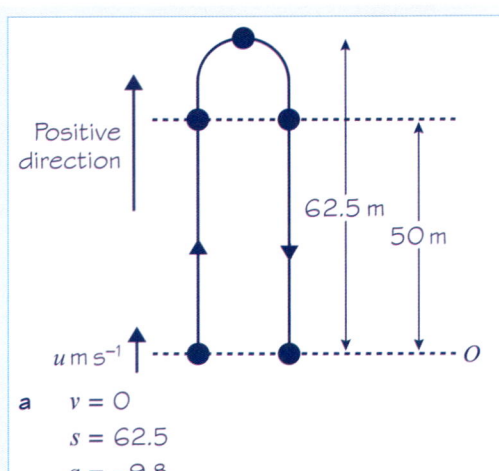

a
$v = 0$
$s = 62.5$
$a = -9.8$
$u = ?$
$v^2 = u^2 + 2as$ — There is no t, so you choose the formula without t.
$0^2 = u^2 + 2 \times (-9.8) \times 62.5$
$u^2 = 1225$
$u = \sqrt{1225} = 35\,\text{m s}^{-1}$ — In this part, you obtain an exact answer, so there is no need for approximation.

The particle will pass through the point 50 m above O twice: once on the way up and once on the way down.

b
$s = 50$
$u = 35$
$a = -9.8$
$t = ?$ — Two values of t need to be found: one on the way up and one on the way down.
$s = ut + \tfrac{1}{2}at^2$
$50 = 35t - 4.9t^2$
$4.9t^2 - 35t + 50 = 0$ — Write the equation in the form $ax^2 + bx + c = 0$ and use the quadratic formula.

$$t = \frac{-b \pm \sqrt{(b^2 - 4ac)}}{2a}$$

$$= \frac{35 \pm \sqrt{(35^2 - 4 \times 4.9 \times 50)}}{9.8}$$

$$= \frac{35 \pm \sqrt{245}}{9.8} \approx \frac{35 \pm 15.6525}{9.8}$$

$t \approx 5.1686\ldots$, or $t \approx 1.9742\ldots$
$(5.1686\ldots) - (1.9742\ldots) \approx 3.194$

Particle is 50 m or more above O for 3.2 s (2 s.f.)

Between these two times the particle is always more than 50 m above O. You find the total time for which the particle is 50 m or more above O by finding the difference of these two values.

CONSTANT ACCELERATION   CHAPTER 2

Example 14   **SKILLS** PROBLEM-SOLVING

A ball A falls vertically from rest at the top of a tower 63 m high. At the same time as A begins to fall, another ball B is projected vertically upward from the bottom of the tower with speed $21\,\mathrm{m\,s^{-1}}$. The balls collide. Find the distance of the point where the balls collide from the bottom of the tower.

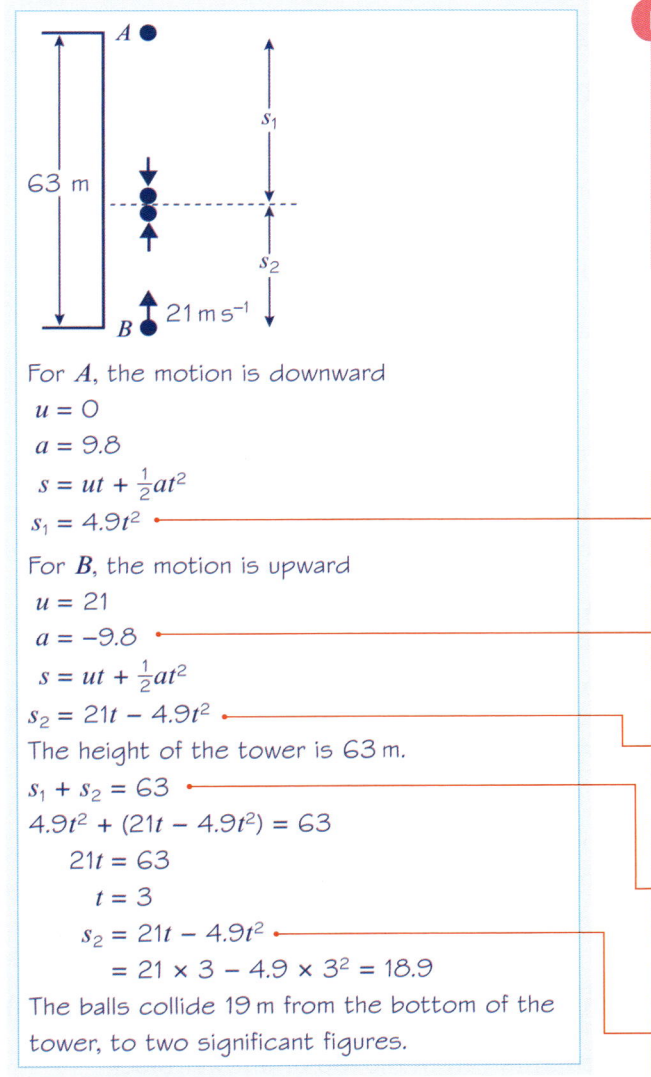

Problem-solving

You must take special care with problems where objects are moving in different directions. Here A is moving downward and you will take the acceleration due to gravity as positive. However, B is moving upward so for B the acceleration due to gravity is negative.

For A, the motion is downward
$u = 0$
$a = 9.8$
$s = ut + \frac{1}{2}at^2$
$s_1 = 4.9t^2$ — You cannot find s_1 at this stage. You have to express it in terms of t.

For B, the motion is upward
$u = 21$
$a = -9.8$ — As B is moving upward, the acceleration due to gravity is negative.
$s = ut + \frac{1}{2}at^2$
$s_2 = 21t - 4.9t^2$ — You now have expressions for s_1 and s_2 in terms of t.

The height of the tower is 63 m.
$s_1 + s_2 = 63$ — Adding together the two distances gives the height of the tower. You can write this as an equation in t.
$4.9t^2 + (21t - 4.9t^2) = 63$
$21t = 63$
$t = 3$
$s_2 = 21t - 4.9t^2$
$= 21 \times 3 - 4.9 \times 3^2 = 18.9$

The balls collide 19 m from the bottom of the tower, to two significant figures.

You have found t but you were asked for the distance from the bottom of the tower. Substitute your value for t into your equation for s_2.

Exercise 2F   **SKILLS** PROBLEM-SOLVING

1 A cliff diver jumps from a point 28 m above the surface of the water. Modelling the diver as a particle moving freely under gravity with initial velocity 0, find:
   a the time taken for the diver to hit the water
   b the speed of the diver when he hits the water.

2 A particle is projected vertically upward with speed $20\,\mathrm{m\,s^{-1}}$ from a point on the ground. Find the time of flight of the particle.

3 A ball is thrown vertically downward from the top of a tower with speed $18\,\text{m\,s}^{-1}$. It reaches the ground in $1.6\,\text{s}$. Find the height of the tower.

4 A pebble is catapulted vertically upward with speed $24\,\text{m\,s}^{-1}$. Find:
   a the greatest height above the point of projection reached by the pebble
   b the time taken to reach this height.

5 A ball is projected upward from a point which is $4\,\text{m}$ above the ground with speed $18\,\text{m\,s}^{-1}$. Find:
   a the speed of the ball when it is $15\,\text{m}$ above its point of projection
   b the speed with which the ball hits the ground.

6 A particle P is projected vertically downward from a point $80\,\text{m}$ above the ground with speed $4\,\text{m\,s}^{-1}$. Find:
   a the speed with which P hits the ground
   b the time P takes to reach the ground.

7 A particle P is projected vertically upward from a point X. Five seconds later, P is moving downward with speed $10\,\text{m\,s}^{-1}$. Find:
   a the speed of projection of P
   b the greatest height above X attained by P during its motion.

8 A ball is thrown vertically upward with speed $21\,\text{m\,s}^{-1}$. It hits the ground $4.5\,\text{s}$ later. Find the height above the ground from which the ball was thrown.

9 A stone is thrown vertically upward from a point which is $3\,\text{m}$ above the ground, with speed $16\,\text{m\,s}^{-1}$. Find:
   a the time of flight of the stone
   b the total distance travelled by the stone.

(P) 10 A particle is projected vertically upward with speed $24.5\,\text{m\,s}^{-1}$. Find the total time for which it is $21\,\text{m}$ or more above its point of projection.

(E/P) 11 A particle is projected vertically upward from a point O with speed $u\,\text{m\,s}^{-1}$. Two seconds later, it is still moving upward and its speed is $\tfrac{1}{3}u\,\text{m\,s}^{-1}$. Find:
   a the value of u **(3 marks)**

   Problem-solving
   Use $v = u + at$ and substitute $v = \tfrac{1}{3}u$.

   b the time from the instant that the particle leaves O to the instant that it returns to O. **(4 marks)**

(E/P) 12 A ball A is thrown vertically downward with speed $5\,\text{m\,s}^{-1}$ from the top of a tower block $46\,\text{m}$ above the ground. At the same time as A is thrown downward, another ball B is thrown vertically upward from the ground with speed $18\,\text{m\,s}^{-1}$. The balls collide. Find the distance of the point where A and B collide from the point where A was thrown. **(5 marks)**

(E/P) 13 A ball is released from rest at a point which is $10\,\text{m}$ above a wooden floor. Each time the ball strikes the floor, it rebounds with three-quarters of the speed with which it strikes the floor. Find the greatest height above the floor reached by the ball:

   Problem-solving
   Consider each bounce as a separate motion.

   a the first time it rebounds from the floor **(3 marks)**
   b the second time it rebounds from the floor. **(4 marks)**

CONSTANT ACCELERATION CHAPTER 2

Challenge

1 A particle P is projected vertically upward from a point O with speed $12\,\text{m s}^{-1}$. One second after P has been projected from O, another particle Q is projected vertically upward from O with speed $20\,\text{m s}^{-1}$. Find: **a** the time between the instant that P is projected from O and the instant when P and Q collide, **b** the distance of the point where P and Q collide from O.

2 A stone is dropped from the top of a building and two seconds later another stone is thrown vertically downward at a speed of $25\,\text{m s}^{-1}$. Both stones reach the ground at the same time. Find the height of the building.

Chapter review 2

1 A car accelerates in a straight line at a constant rate, starting from rest at a point A and reaching a velocity of $45\,\text{km h}^{-1}$ in $20\,\text{s}$. This velocity is then maintained and the car passes a point B 3 minutes after leaving A.
   a Sketch a velocity–time graph to illustrate the motion of the car.
   b Find the displacement of the car from its starting point after 3 minutes.

2 A particle is moving on an axis Ox. From time $t = 0$ to time $t = 32\,\text{s}$, the particle is travelling with constant velocity $15\,\text{m s}^{-1}$. The particle then decelerates from $15\,\text{m s}^{-1}$ to rest in T seconds.
   a Sketch a velocity–time graph to illustrate the motion of the particle.
   The total distance travelled by the particle is $570\,\text{m}$.
   b Find the value of T.
   c Sketch a displacement–time graph illustrating the motion of the particle.

(P) 3 The velocity–time graph represents the motion of a particle moving in a straight line accelerating from velocity u at time 0 to velocity v at time t.
   a Use the graph to show that:
   i $v = u + at$ **ii** $s = \left(\dfrac{u+v}{2}\right)t$
   b Hence show that:
   i $v^2 = u^2 + 2as$ **ii** $s = ut + \tfrac{1}{2}at^2$ **iii** $s = vt - \tfrac{1}{2}at^2$

(P) 4 The diagram is a velocity–time graph representing the motion of a cyclist along a straight road. At time $t = 0$, the cyclist is moving with velocity $u\,\text{m s}^{-1}$. The velocity is maintained until time $t = 15\,\text{s}$, when she slows down with constant deceleration, coming to rest when $t = 23\,\text{s}$. The total distance she travels in 23 seconds is $152\,\text{m}$. Find the value of u.

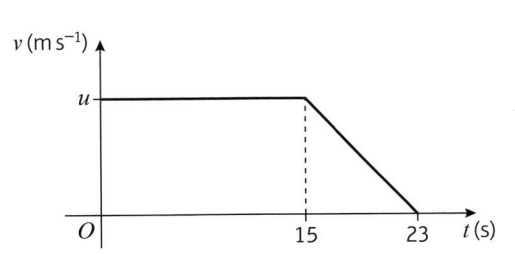

5 At $t = 0$, a car is travelling at $2\,\text{m}\,\text{s}^{-1}$. It then moves with constant acceleration of $1\,\text{m}\,\text{s}^{-2}$ for $4\,\text{s}$. It then moves with constant velocity for $4\,\text{s}$ before decelerating at $2\,\text{m}\,\text{s}^{-2}$ for $2\,\text{s}$. Draw an acceleration-time graph to represent the motion of the car.

6 A car travelling on a straight road slows down with constant deceleration. The car is travelling at $40\,\text{km}\,\text{h}^{-1}$ as it passes a post box, and at $24\,\text{km}\,\text{h}^{-1}$ as it passes a road sign. The distance between the road sign and the post box is $240\,\text{m}$. Find, in $\text{m}\,\text{s}^{-2}$, the deceleration of the car.

7 A particle P is moving along the x-axis with constant deceleration $2.5\,\text{m}\,\text{s}^{-2}$. At time $t = 0$, P passes through the origin O with velocity $20\,\text{m}\,\text{s}^{-1}$ in the direction of x increasing. At time $t = 12\,\text{s}$, P is at the point A. Find:
  a the distance OA **b** the total distance P travels in $12\,\text{s}$.

8 A ball is thrown vertically downward from the top of a tower with speed $6\,\text{m}\,\text{s}^{-1}$. The ball strikes the ground with speed $25\,\text{m}\,\text{s}^{-1}$. Find the time the ball takes to travel from the top of the tower to the ground.

9 A child drops a ball from a point at the top of a cliff which is $82\,\text{m}$ above the sea. The ball is initially **at rest**.
  a Find:
    i the time taken for the ball to reach the sea   **ii** the speed with which the ball hits the sea.
  b State one physical factor which has been ignored in making your calculation.

(P) 10 A particle moves $451\,\text{m}$ in a straight line. The diagram shows a speed–time graph illustrating the motion of the particle. The particle starts at rest and accelerates at a constant rate for $8\,\text{s}$ reaching a speed of $2u\,\text{m}\,\text{s}^{-1}$. The particle then travels at a constant speed for 12 seconds before decelerating uniformly, reaching a speed of $u\,\text{m}\,\text{s}^{-1}$ at time $t = 26\,\text{s}$. Find:
  a the value of u
  b the distance travelled by the particle while its speed is less than $u\,\text{m}\,\text{s}^{-1}$.

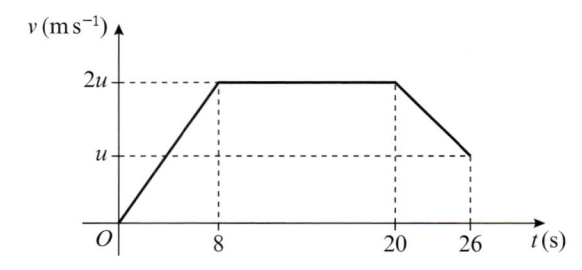

(E/P) 11 A train is travelling with constant acceleration along a straight track. At time $t = 0$, the train passes a point O, travelling with velocity $18\,\text{m}\,\text{s}^{-1}$. At time $t = 12\,\text{s}$, the train passes a point P, travelling with velocity $24\,\text{m}\,\text{s}^{-1}$. At time $t = 20\,\text{s}$, the train passes a point Q. Find:
  a the speed of the train at Q **(5 marks)**
  b the distance from P to Q.   **(2 marks)**

(E) 12 A particle moves along a straight line, from a point X to a point Y, with constant acceleration. The distance from X to Y is $104\,\text{m}$. The particle takes $8\,\text{s}$ to move from X to Y and the speed of the particle at Y is $18\,\text{m}\,\text{s}^{-1}$. Find:
  a the speed of the particle at X **(3 marks)**
  b the acceleration of the particle.   **(2 marks)**
The particle continues to move with the same acceleration until it reaches a point Z. At Z the speed of the particle is three times the speed of the particle at X.
  c Find the distance XZ.   **(4 marks)**

CONSTANT ACCELERATION   CHAPTER 2   37

(E) 13 A pebble is projected vertically upward with speed 21 m s^{-1} from a point 32 m above the ground. Find:
    a the speed with which the pebble strikes the ground **(3 marks)**
    b the total time for which the pebble is more than 40 m above the ground. **(4 marks)**
    c Sketch a velocity–time graph for the motion of the pebble from the instant it is projected to the instant it hits the ground, showing the values of t at any points where the graph intercepts the horizontal axis. **(4 marks)**

(E) 14 A car is moving along a straight road with uniform acceleration. The car passes a checkpoint A with speed 12 m s^{-1} and another checkpoint C with speed 32 m s^{-1}. The distance between A and C is 1100 m.
    a Find the time taken by the car to move from A to C. **(2 marks)**
    b Given that B is the midpoint of AC, find the speed with which the car passes B. **(2 marks)**

(E/P) 15 A particle is projected vertically upward with a speed of 30 m s^{-1} from a point A. The point B is h metres above A. The particle moves freely under gravity and is above B for a time 2.4 s. Calculate the value of h. **(5 marks)**

(E/P) 16 Two cars A and B are moving in the same direction along a straight horizontal road. At time $t = 0$, they are side by side, passing a point O on the road. Car A travels at a constant speed of 30 m s^{-1}. Car B passes O with a speed of 20 m s^{-1}, and has constant acceleration of 4 m s^{-2}. Find:
    a the speed of B when it has travelled 78 m from O **(2 marks)**
    b the distance from O of A when B is 78 m from O **(3 marks)**
    c the time when B overtakes A. **(4 marks)**

(E/P) 17 A car is being driven on a straight stretch of motorway at a constant velocity of 34 m s^{-1} when it passes a velocity restriction sign S warning of road works ahead and requiring speeds to be reduced to 22 m s^{-1}. The driver continues at her velocity for 2 s after passing S. She then reduces her velocity to 22 m s^{-1} with constant deceleration of 3 m s^{-2}, and continues at the lower velocity.
    a Draw a velocity–time graph to illustrate the motion of the car after it passes S. **(2 marks)**
    b Find the shortest distance before the road works that S should be placed on the road to ensure that a car driven in this way has had its velocity reduced to 22 m s^{-1} by the time it reaches the start of the road works. **(4 marks)**

(E/P) 18 A train starts from rest at station A and accelerates uniformly at $3x \text{ m s}^{-2}$ until it reaches a velocity of 30 m s^{-1}. For the next T seconds the train maintains this constant velocity. The train then decelerates uniformly at $x \text{ m s}^{-2}$ until it comes to rest at a station B. The distance between the stations is 6 km and the time taken from A to B is 5 minutes.
    a Sketch a velocity–time graph to illustrate this journey. **(2 marks)**
    b Show that $\frac{40}{x} + T = 300$. **(4 marks)**
    c Find the value of T and the value of x. **(2 marks)**
    d Calculate the distance the train travels at constant velocity. **(2 marks)**
    e Calculate the time taken from leaving A until reaching the point halfway between the stations. **(3 marks)**

Challenge

A ball is projected vertically upward with speed 10 m s⁻¹ from a point X, which is 50 m above the ground. T seconds after the first ball is projected upward, a second ball is dropped from X. Initially the second ball is at rest. The balls collide 25 m above the ground. Find the value of T.

Summary of key points

1. Velocity is the **rate of change** of displacement.

    On a displacement–time graph the **gradient** represents the velocity.

    If the displacement–time graph is a straight line, then the velocity is constant.

2. Average velocity = $\dfrac{\text{displacement from starting point}}{\text{time taken}}$

3. Average speed = $\dfrac{\text{total distance travelled}}{\text{time taken}}$

4. Acceleration is the **rate of change** of velocity.

    In a velocity–time graph the **gradient** represents the acceleration.

    If the velocity–time graph is a straight line, then the acceleration is constant.

5. The area between a velocity–time graph and the horizontal axis represents the distance travelled.

    For motion in a straight line with positive velocity, the area under the velocity–time graph up to a point t represents the displacement at time t.

6. The area between the acceleration-time graph and the horizontal axis represents the change in velocity. In other words, the area under the acceleration-time graph for a certain time interval is equal to the change in velocity during that time interval.

7. You need to be able to use and recall the five formulae for solving problems about particles moving in a straight line with constant acceleration.

    - $v = u + at$
    - $s = \left(\dfrac{u + v}{2}\right)t$
    - $v^2 = u^2 + 2as$
    - $s = ut + \tfrac{1}{2}at^2$
    - $s = vt - \tfrac{1}{2}at^2$

8. The force of **gravity** causes all objects to accelerate toward the Earth. If you ignore the effects of air resistance, this acceleration is constant. It does not depend on the mass of the object.

9. An object moving vertically in a straight line can be modelled as a particle with a constant downward acceleration of $g = 9.8$ m s⁻².

3 VECTORS IN MECHANICS

2.1
2.2

Learning objectives

After completing this chapter you should be able to:

- Use vectors in two dimensions → pages 40–42
- Calculate the magnitude and direction of a vector → pages 43–44
- Understand and use position vectors → pages 46–49
- Understand vector magnitude and use vectors in speed and distance calculations → pages 46–49
- Use vectors to solve problems in context → pages 46–49

Prior knowledge check

1 Write the column vector for the translation of shape:

   a A to B
   b A to C
   c A to D

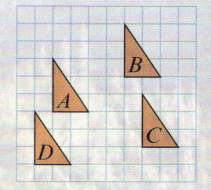

   ← International GCSE Mathematics

2 Find x to one decimal place:

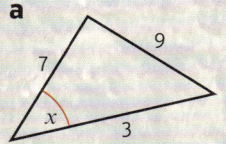

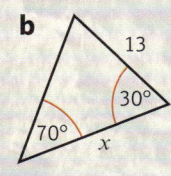

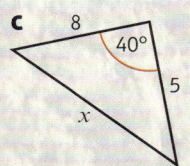

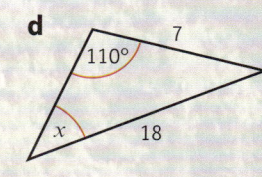

   a, **b**, **c**, **d**

   ← International GCSE Mathematics

3 Given that $\overrightarrow{OA} = 3\mathbf{a}$ and $\overrightarrow{OB} = 4\mathbf{b}$, find vector $\overrightarrow{AB}$.
   Hence find the magnitude of $\overrightarrow{AB}$.

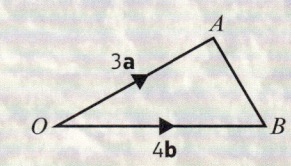

   ← International GCSE Mathematics

Pilots use vector addition to work out the resultant vector for their speed and heading when an aeroplane encounters a strong cross-wind. Engineers also use vectors to work out the resultant forces acting on buildings in construction.

3.1 Working with vectors

- A vector is a quantity which has both magnitude and direction.
  These are examples of **vector** quantities:

Quantity	Description	Unit
Displacement	distance in a particular direction	metre (m)
Velocity	**rate of change** of displacement	metres per second (m s^{-1})
Acceleration	rate of change of velocity	metres per second per second (m s^{-2})
Force/weight	described by magnitude, direction and point of application	newton (N)

- A scalar quantity has magnitude only.
  These are examples of **scalar** quantities:

Quantity	Description	Unit
Distance	measure of length	metre (m)
Speed	measure of how quickly a body moves	metres per second (m s^{-1})
Time	measure of ongoing events taking place	second (s)
Mass	measure of the quantity of matter contained in an object	kilogram (kg)

Scalar quantities are always **positive**. When considering motion in a straight line (1-dimensional), **vector** quantities can be **positive** or **negative**.

Example 1   SKILLS   PROBLEM-SOLVING

A girl walks 2 km due east from a fixed point O to A, and then 3 km due south from A to B. Find the total distance walked and describe the displacement of B from O.

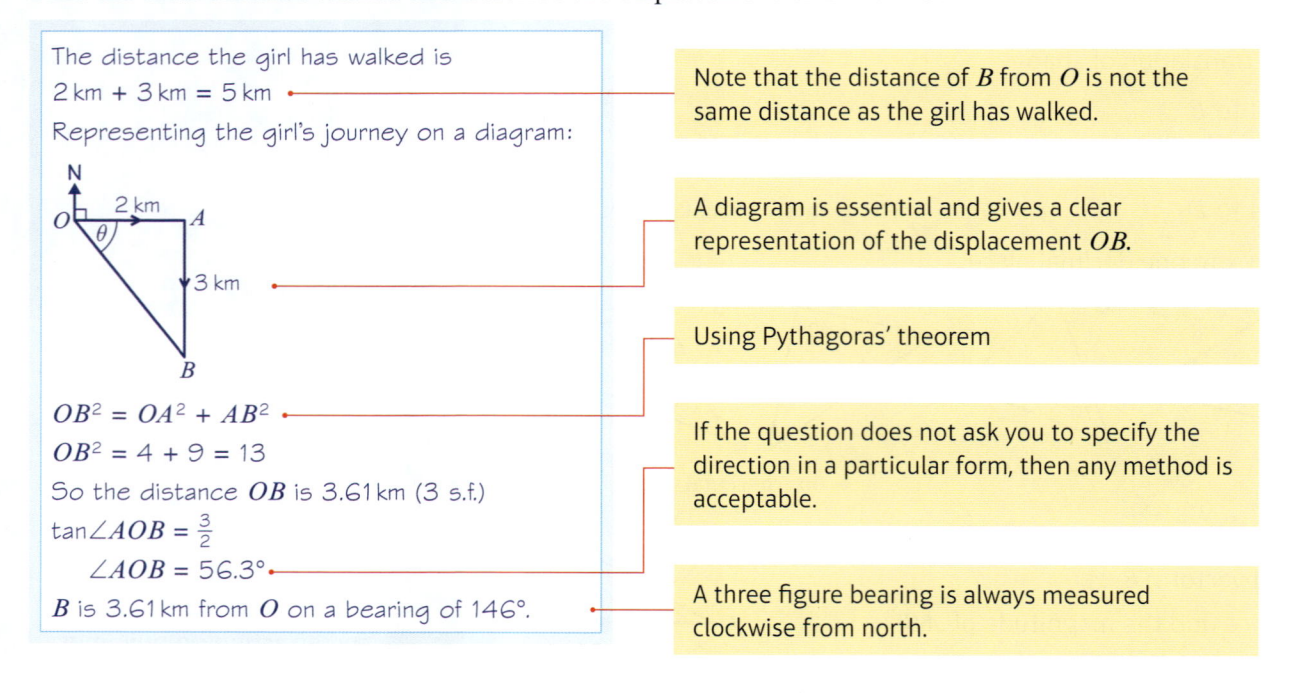

The distance the girl has walked is
2 km + 3 km = 5 km
Representing the girl's journey on a diagram:

$OB^2 = OA^2 + AB^2$
$OB^2 = 4 + 9 = 13$
So the distance OB is 3.61 km (3 s.f.)
$\tan \angle AOB = \frac{3}{2}$
$\angle AOB = 56.3°$
B is 3.61 km from O on a bearing of 146°.

— Note that the distance of B from O is not the same distance as the girl has walked.

— A diagram is essential and gives a clear representation of the displacement OB.

— Using Pythagoras' theorem

— If the question does not ask you to specify the direction in a particular form, then any method is acceptable.

— A three figure bearing is always measured clockwise from north.

VECTORS IN MECHANICS — CHAPTER 3

You can describe vectors using **i** and **j** notation.

- **A unit vector is a vector of length 1. The unit vectors along the Cartesian axes are usually denoted by i and j respectively.**

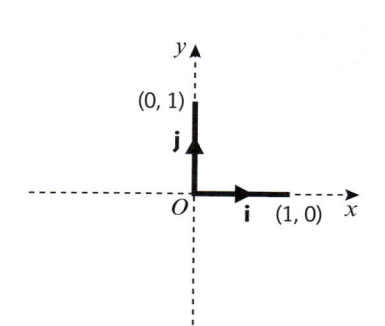

- **You can write any two-dimensional vector in the form $a\mathbf{i} + b\mathbf{j}$.**

By the triangle law of addition:
$$\overrightarrow{AC} = \overrightarrow{AB} + \overrightarrow{BC}$$
$$= 5\mathbf{i} + 2\mathbf{j}$$

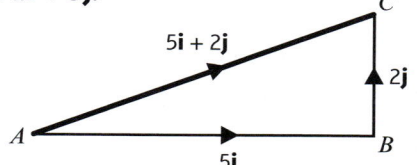

Example 2

Draw a diagram to represent the vector $-3\mathbf{i} + \mathbf{j}$.

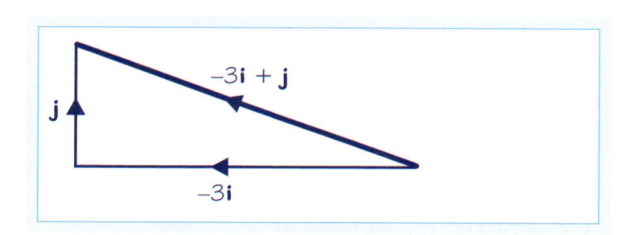

3 units in the direction of the unit vector $-\mathbf{i}$ and 1 unit in the direction of the unit vector $\mathbf{j}$.

Example 3

The diagram shows vectors **a**, **b** and **c**. Draw a diagram to illustrate the vector addition $\mathbf{a} + \mathbf{b} + \mathbf{c}$.

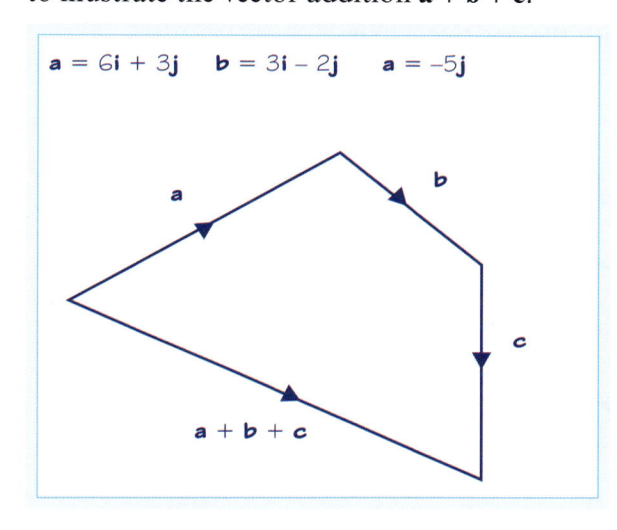

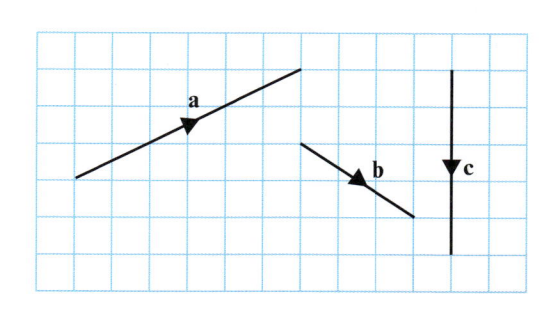

$\mathbf{a} = 6\mathbf{i} + 3\mathbf{j} \quad \mathbf{b} = 3\mathbf{i} - 2\mathbf{j} \quad \mathbf{a} = -5\mathbf{j}$

Add the vectors together and collect up **i** and **j** components.

$\mathbf{a} + \mathbf{b} + \mathbf{c} = 6\mathbf{i} + 3\mathbf{j} + 3\mathbf{i} - 2\mathbf{j} - 5\mathbf{j} = 9\mathbf{i} - 4\mathbf{j}$

Exercise 3A SKILLS PROBLEM-SOLVING

1. A bird flies 5 km due north and then 7 km due east. How far is the bird from its original position, and in what direction?

2. A girl cycles 4 km due west and then 6 km due north. Calculate the total distance she has cycled, and her displacement from her starting point.

3. **a** Express the vectors v_1, v_2, v_3, v_4, v_5 and v_6 using the **i** and **j** notation.
   b Find:
   i $v_1 + v_2$
   ii $v_4 + v_5$
   iii $v_6 + v_1 + v_5$

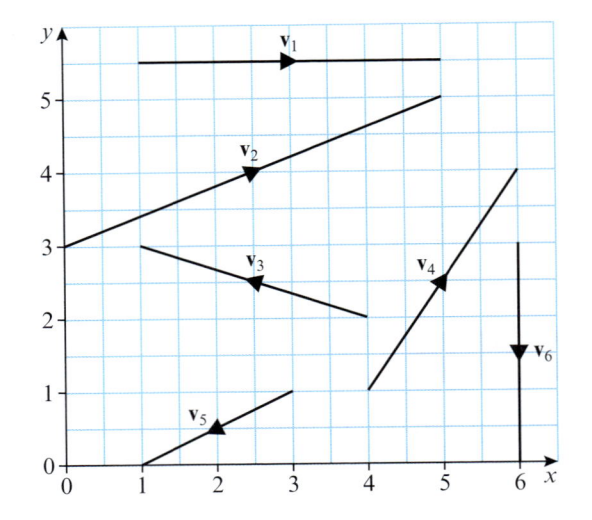

4. Find the magnitude of each of these vectors:
   a $3\mathbf{i} + 4\mathbf{j}$
   b $6\mathbf{i} - 8\mathbf{j}$
   c $5\mathbf{i} + 12\mathbf{j}$
   d $2\mathbf{i} + 4\mathbf{j}$
   e $3\mathbf{i} - 5\mathbf{j}$
   f $4\mathbf{i} + 7\mathbf{j}$
   g $-3\mathbf{i} + 5\mathbf{j}$
   h $-4\mathbf{i} - \mathbf{j}$

5. $\mathbf{a} = 2\mathbf{i} + 3\mathbf{j}$, $\mathbf{b} = 3\mathbf{i} - 4\mathbf{j}$ and $\mathbf{c} = 5\mathbf{i} - \mathbf{j}$. Find the exact value of the magnitude of:
   a $\mathbf{a} + \mathbf{b}$
   b $2\mathbf{a} - \mathbf{c}$
   c $3\mathbf{b} - 2\mathbf{c}$

6. Find the angle that each of these vectors makes with the positive x-axis:
   a $3\mathbf{i} + 4\mathbf{j}$
   b $6\mathbf{i} - 8\mathbf{j}$
   c $5\mathbf{i} + 12\mathbf{j}$
   d $2\mathbf{i} + 4\mathbf{j}$

7. Find the angle that each of these vectors makes with the positive y-axis:
   a $3\mathbf{i} - 5\mathbf{j}$
   b $4\mathbf{i} + 7\mathbf{j}$
   c $-3\mathbf{i} + 5\mathbf{j}$
   d $-4\mathbf{i} - \mathbf{j}$

Challenge

In the diagram, $\overrightarrow{AB} = p\mathbf{i} + q\mathbf{j}$ and $\overrightarrow{AD} = r\mathbf{i} + s\mathbf{j}$. $ABCD$ is a parallelogram.

Prove that the area of $ABCD$ is $ps - qr$.

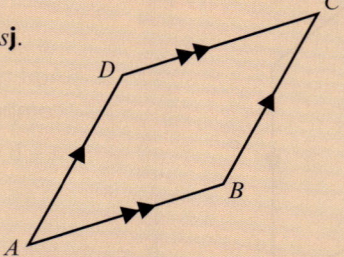

Problem-solving

Draw the parallelogram on a coordinate grid, and choose a position for the origin that will simplify your calculations.

VECTORS IN MECHANICS   CHAPTER 3

3.2 Solving problems with vectors written using **i** and **j** notation

- When vectors are written in terms of the unit vectors **i** and **j** you can add them together by adding the terms in **i** and **j** separately. You can subtract vectors in a similar way.

Example 4

Given **p** = 2**i** + 3**j** and **q** = 5**i** + **j**, find **p** + **q** in terms of **i** and **j**.

$$\begin{aligned} \mathbf{p} + \mathbf{q} &= (2\mathbf{i} + 3\mathbf{j}) + (5\mathbf{i} + \mathbf{j}) \\ &= (2\mathbf{i} + 5\mathbf{i}) + (3\mathbf{j} + \mathbf{j}) \\ &= 7\mathbf{i} + 4\mathbf{j} \end{aligned}$$

Rearrange with the **i** terms together and the **j** terms together.

Simplify the answer.

Example 5

Given **a** = 5**i** + 2**j** and **b** = 3**i** − 4**j**, find 2**a** − **b** in terms of **i** and **j**.

$$\begin{aligned} 2\mathbf{a} &= 2(5\mathbf{i} + 2\mathbf{j}) \\ &= 10\mathbf{i} + 4\mathbf{j} \\ 2\mathbf{a} - \mathbf{b} &= (10\mathbf{i} + 4\mathbf{j}) - (3\mathbf{i} - 4\mathbf{j}) \\ &= (10\mathbf{i} - 3\mathbf{i}) + (4\mathbf{j} - (-4\mathbf{j})) \\ &= 7\mathbf{i} + 8\mathbf{j} \end{aligned}$$

When you double **a**, the **i** term and the **j** term are both doubled – just the same as when you multiply out a bracket in algebra.

Rearrange with the **i** terms together and the **j** terms together.

Simplify the answer but take care when subtracting a negative term.

- When a vector is written in terms of the unit vectors **i** and **j** you can find the magnitude using Pythagoras' Theorem. The magnitude of vector **a** is written |**a**|.

Example 6

Given **v** = 3**i** − 7**j**, find the magnitude of **v**.

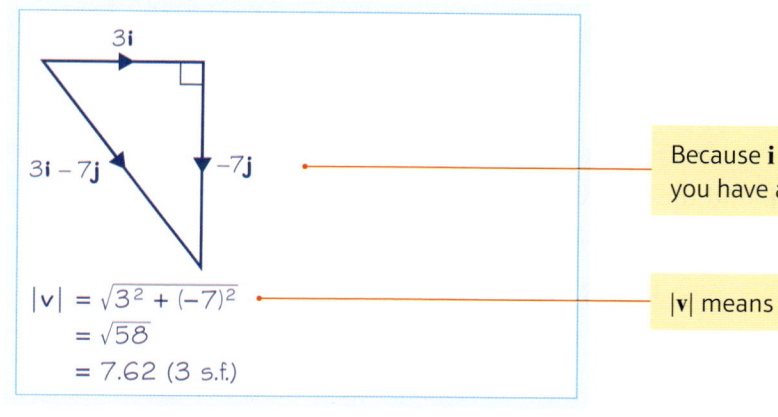

$$\begin{aligned} |\mathbf{v}| &= \sqrt{3^2 + (-7)^2} \\ &= \sqrt{58} \\ &= 7.62 \ (3 \ \text{s.f.}) \end{aligned}$$

Because **i** and **j** are perpendicular vectors, you have a right-angled triangle.

|**v**| means the magnitude of **v**.

Example 7

Find the angle between the vector $4\mathbf{i} + 5\mathbf{j}$ and the positive x-axis.

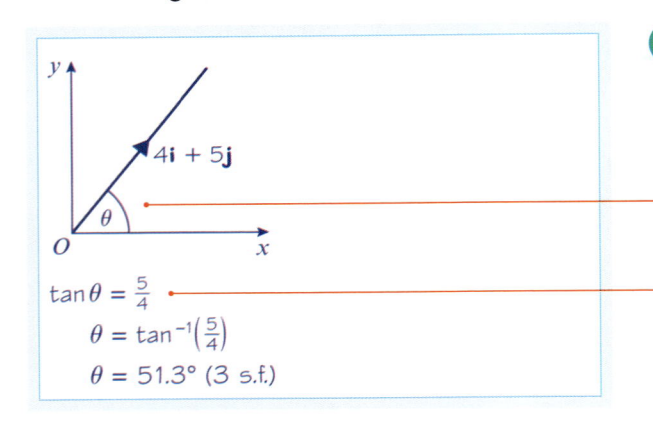

Hint Always draw a diagram first.

[Diagram: vector $4\mathbf{i} + 5\mathbf{j}$ in the first quadrant, with angle θ measured from positive x-axis.]

— Identify the angle you need to find.

$\tan\theta = \frac{5}{4}$

$\theta = \tan^{-1}\left(\frac{5}{4}\right)$

$\theta = 51.3°$ (3 s.f.)

— We have a right-angled triangle, so finding the angle is straightforward.

Example 8

Given that $\mathbf{p} = 4\mathbf{i} + 3\mathbf{j}$ and $\mathbf{q} = 2\mathbf{i} - 6\mathbf{j}$, find λ if $\mathbf{p} + \lambda\mathbf{q}$ is parallel to the vector:

a $\mathbf{j}$ **b** $\mathbf{i} + \mathbf{j}$

a $\mathbf{p} = 4\mathbf{i} + 3\mathbf{j}, \mathbf{p} = 2\mathbf{i} - 6\mathbf{j}$

$\mathbf{p} + \lambda\mathbf{q} = 4\mathbf{i} + 3\mathbf{j} + \lambda(2\mathbf{i} + 6\mathbf{j})$

$= 4\mathbf{i} + 2\lambda\mathbf{i} + 3\mathbf{j} - 6\lambda\mathbf{j}$

$4\mathbf{i} + 2\lambda\mathbf{i} = 0$

$2\lambda\mathbf{i} = -4\mathbf{i}$

$\Rightarrow \lambda = -2$

— Add together $\mathbf{p}$ and $\lambda\mathbf{q}$.

— $\mathbf{p} + \lambda\mathbf{q}$ is parallel to $\mathbf{j}$, therefore the $\mathbf{i}$-component of $\mathbf{p} + \lambda\mathbf{q}$ is 0.

b $\mathbf{p} = 4\mathbf{i} + 3\mathbf{j}$

$\mathbf{q} = 2\mathbf{i} - 6\mathbf{j} \Rightarrow \lambda\mathbf{q} = 2\lambda\mathbf{i} - 6\lambda\mathbf{j}$

$\mathbf{p} + \lambda\mathbf{q} = (4\mathbf{i} + 3\mathbf{j}) + (2\lambda\mathbf{i} - 6\lambda\mathbf{j})$

$= \mathbf{i}(4 + 2\lambda) + \mathbf{j}(3 - 6\lambda)$

$4 + 2\lambda = 3 - 6\lambda$

$\Rightarrow \lambda = -\frac{1}{8}$

— We require the vector in the form $\mathbf{p} + \lambda\mathbf{q}$ so multiply $\mathbf{q}$ through by λ.

— Rearrange with the $\mathbf{i}$ terms together and the $\mathbf{j}$ terms together.

— The $\mathbf{i}$ and $\mathbf{j}$ components must be equal so equate them and solve for λ.

Exercise 3B

1 Given that $\mathbf{a} = 2\mathbf{i} + 3\mathbf{j}$ and $\mathbf{b} = 4\mathbf{i} - \mathbf{j}$, find these in terms of $\mathbf{i}$ and $\mathbf{j}$.

a $\mathbf{a} + \mathbf{b}$ **b** $3\mathbf{a} + \mathbf{b}$ **c** $2\mathbf{a} - \mathbf{b}$ **d** $2\mathbf{b} - \mathbf{a}$

e $3\mathbf{a} - 2\mathbf{b}$ **f** $\mathbf{b} - 3\mathbf{a}$ **g** $4\mathbf{b} - \mathbf{a}$ **h** $2\mathbf{a} - 3\mathbf{b}$

2 Given that $\mathbf{a} = 2\mathbf{i} + 5\mathbf{j}$ and $\mathbf{b} = 3\mathbf{i} - \mathbf{j}$, find:

a λ if $\mathbf{a} + \lambda\mathbf{b}$ is parallel to the vector $\mathbf{i}$

b μ if $\mu\mathbf{a} + \mathbf{b}$ is parallel to the vector $\mathbf{j}$.

3 Given that $c = 3i + 4j$ and $d = i - 2j$, find:

  a λ if $c + \lambda d$ is parallel to $i + j$
  b μ if $\mu c + d$ is parallel to $i + 3j$
  c s if $c - sd$ is parallel to $2i + j$
  d t if $d - tc$ is parallel to $-2i + 3j$.

4 In this question, the horizontal unit vectors i and j are directed due east and due north respectively. Find the magnitude and bearing of these vectors:

  a $2i + 3j$ **b** $4i - j$ **c** $-3i + 2j$ **d** $-2i - j$

Challenge

SKILLS ADAPTIVE LEARNING

Given that $a = 3i - 5j$ and $b = i + 4j$, find the values of λ if $a + \lambda b$ is perpendicular to $3i - 2j$.
Comment on your answer.

3.3 The velocity of a particle as a vector

- The velocity of a particle is a vector in the direction of motion. Its magnitude is the speed of the particle. The velocity is usually denoted by **v**.

If a particle is moving with constant velocity v m s^{-1}, then after time t seconds it will have moved vt m. The displacement is parallel to the velocity. The magnitude of the displacement is the distance from the starting point.

Example 9

A particle is moving with constant velocity $v = (3i + j)$ m s^{-1}. Find **a** the speed of the particle, **b** the distance moved every 4 seconds.

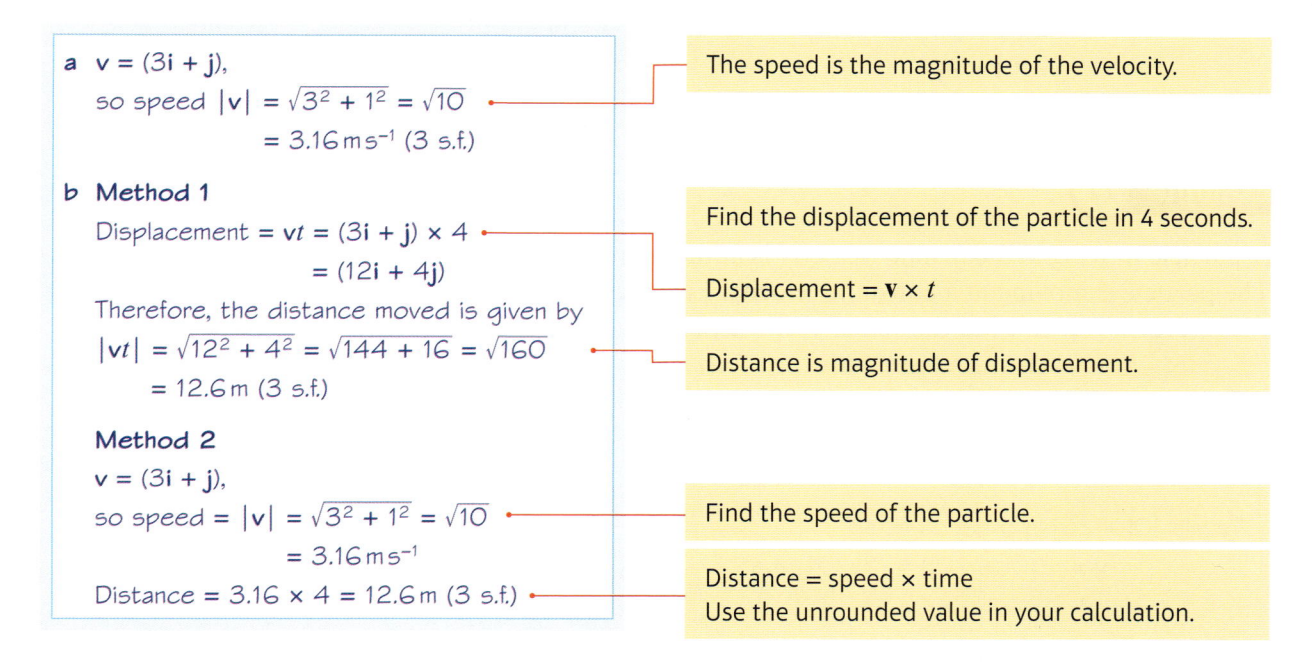

a $v = (3i + j)$,
so speed $|v| = \sqrt{3^2 + 1^2} = \sqrt{10}$
$= 3.16$ m s^{-1} (3 s.f.)

The speed is the magnitude of the velocity.

b Method 1
Displacement $= vt = (3i + j) \times 4$
$= (12i + 4j)$

Find the displacement of the particle in 4 seconds.

Displacement $= v \times t$

Therefore, the distance moved is given by
$|vt| = \sqrt{12^2 + 4^2} = \sqrt{144 + 16} = \sqrt{160}$
$= 12.6$ m (3 s.f.)

Distance is magnitude of displacement.

Method 2
$v = (3i + j)$,
so speed $= |v| = \sqrt{3^2 + 1^2} = \sqrt{10}$
$= 3.16$ m s^{-1}

Find the speed of the particle.

Distance $= 3.16 \times 4 = 12.6$ m (3 s.f.)

Distance = speed × time
Use the unrounded value in your calculation.

Exercise 3C SKILLS EXECUTIVE FUNCTION

1 Find the speed of a particle moving with these velocities:
  a $(3\mathbf{i} + 4\mathbf{j})\,\text{m s}^{-1}$
  b $(24\mathbf{i} - 7\mathbf{j})\,\text{km h}^{-1}$
  c $(5\mathbf{i} + 2\mathbf{j})\,\text{m s}^{-1}$
  d $(-7\mathbf{i} + 4\mathbf{j})\,\text{cm s}^{-1}$

2 Find the distance moved by a particle which travels for:
  a 5 hours at velocity $(8\mathbf{i} + 6\mathbf{j})\,\text{km h}^{-1}$
  b 10 seconds at velocity $(5\mathbf{i} - \mathbf{j})\,\text{m s}^{-1}$
  c 45 minutes at velocity $(6\mathbf{i} + 2\mathbf{j})\,\text{km h}^{-1}$
  d 2 minutes at velocity $(-4\mathbf{i} - 7\mathbf{j})\,\text{cm s}^{-1}$

3 Find the speed and the distance travelled by a particle moving with:
  a velocity $(-3\mathbf{i} + 4\mathbf{j})\,\text{m s}^{-1}$ for 15 seconds
  b velocity $(2\mathbf{i} + 5\mathbf{j})\,\text{m s}^{-1}$ for 3 seconds
  c velocity $(5\mathbf{i} - 2\mathbf{j})\,\text{km h}^{-1}$ for 3 hours
  d velocity $(12\mathbf{i} - 5\mathbf{j})\,\text{km h}^{-1}$ for 30 minutes

3.4 Solving problems involving velocity and time using vectors

- If a particle starts from the point with position vector $\mathbf{r}_0$ and moves with constant velocity $\mathbf{v}$, then its displacement from its initial position at time t is $\mathbf{v}t$ and its position vector $\mathbf{r}$ is given by:

  $\mathbf{r} = \mathbf{r}_0 + \mathbf{v}t$

Example 10 SKILLS PROBLEM-SOLVING

A particle starts from the point with position vector $(3\mathbf{i} + 7\mathbf{j})\,\text{m}$ and moves with constant velocity $(2\mathbf{i} - \mathbf{j})\,\text{m s}^{-1}$. Find the position vector of the particle 4 seconds later.

Displacement $= \mathbf{v}t = 4(2\mathbf{i} - \mathbf{j}) = 8\mathbf{i} - 4\mathbf{j}$ ← Distance $= \mathbf{v} \times t$
Position vector $\mathbf{r} = (3\mathbf{i} + 7\mathbf{j}) + (8\mathbf{i} - 4\mathbf{j})$ ← $\mathbf{r} = \mathbf{r}_0 + \mathbf{v}t$
$= (3 + 8)\mathbf{i} + (7 - 4)\mathbf{j}$
$= 11\mathbf{i} + 3\mathbf{j}$ ← Position vector after 4 seconds = position vector of starting point + displacement.

Example 11

A particle moving at a constant velocity is at the point with position vector $(2\mathbf{i} - 4\mathbf{j})\,\text{m}$ at time $t = 0$. The particle is moving at a constant velocity. Five seconds later it is at the point with position vector $(12\mathbf{i} + 16\mathbf{j})\,\text{m}$. Find the velocity of the particle.

Displacement $= (12\mathbf{i} + 16\mathbf{j}) - (2\mathbf{i} - 4\mathbf{j})$ ← You need to use the same formula $\mathbf{r} = \mathbf{r}_0 + \mathbf{v}t$, but this time you are using it to find $\mathbf{v}$.
$= (12 - 2)\mathbf{i} + (16 + 4)\mathbf{j}$
$= 10\mathbf{i} + 20\mathbf{j}$ ← Start by finding the displacement.
Travels $10\mathbf{i} + 20\mathbf{j}$ in 5 seconds, so
$\mathbf{v} = \tfrac{1}{5}(10\mathbf{i} + 20\mathbf{j}) = (2\mathbf{i} + 4\mathbf{j})\,\text{m s}^{-1}$ ← Travelling at constant velocity, so divide displacement by time taken to obtain $\mathbf{v}$.

VECTORS IN MECHANICS — CHAPTER 3

Example 12

At time $t = 0$, a particle has position vector $4\mathbf{i} + 7\mathbf{j}$ and is moving with speed $15\,\text{m s}^{-1}$ in the direction $3\mathbf{i} - 4\mathbf{j}$. Find its position vector after 2 seconds.

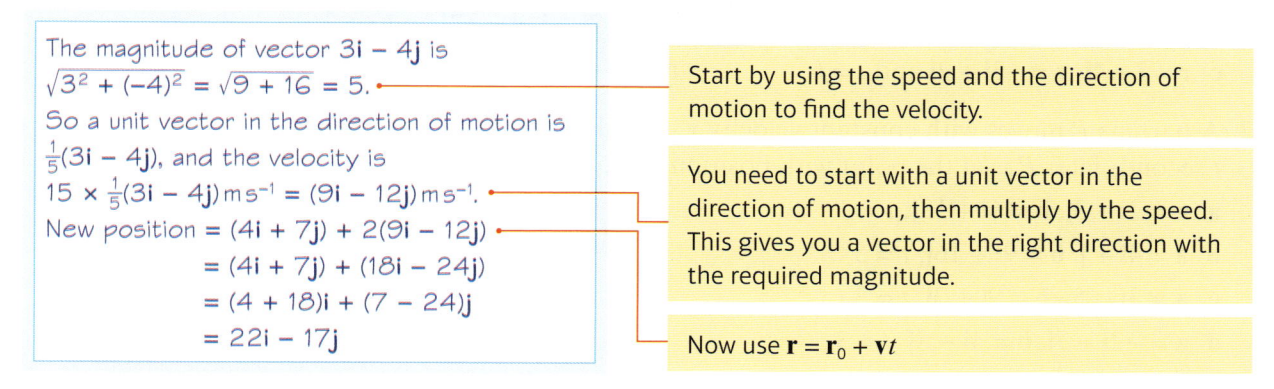

> The magnitude of vector $3\mathbf{i} - 4\mathbf{j}$ is
> $\sqrt{3^2 + (-4)^2} = \sqrt{9 + 16} = 5$.
> So a unit vector in the direction of motion is
> $\frac{1}{5}(3\mathbf{i} - 4\mathbf{j})$, and the velocity is
> $15 \times \frac{1}{5}(3\mathbf{i} - 4\mathbf{j})\,\text{m s}^{-1} = (9\mathbf{i} - 12\mathbf{j})\,\text{m s}^{-1}$.
> New position $= (4\mathbf{i} + 7\mathbf{j}) + 2(9\mathbf{i} - 12\mathbf{j})$
> $= (4\mathbf{i} + 7\mathbf{j}) + (18\mathbf{i} - 24\mathbf{j})$
> $= (4 + 18)\mathbf{i} + (7 - 24)\mathbf{j}$
> $= 22\mathbf{i} - 17\mathbf{j}$

Start by using the speed and the direction of motion to find the velocity.

You need to start with a unit vector in the direction of motion, then multiply by the speed. This gives you a vector in the right direction with the required magnitude.

Now use $\mathbf{r} = \mathbf{r}_0 + \mathbf{v}t$

- The acceleration of a particle tells you how the velocity changes with time. Acceleration is a vector, usually denoted by $\mathbf{a}$. If a particle with initial velocity $\mathbf{u}$ moves with constant acceleration $\mathbf{a}$ then its velocity, $\mathbf{v}$ at time t is given by:

  $\mathbf{v} = \mathbf{u} + \mathbf{a}t$

Example 13

A particle P has velocity $(-3\mathbf{i} + \mathbf{j})\,\text{m s}^{-1}$ at time $t = 0$. The particle moves with constant acceleration $\mathbf{a} = (2\mathbf{i} + 3\mathbf{j})\,\text{m s}^{-2}$. Find the speed of the particle after 3 seconds.

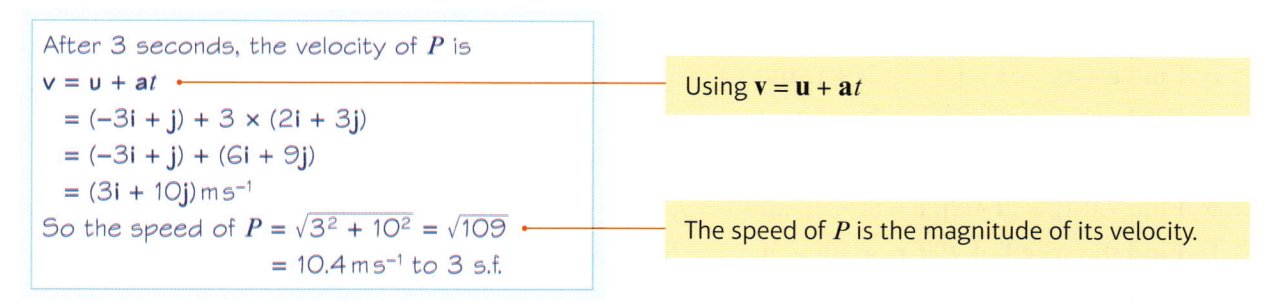

> After 3 seconds, the velocity of P is
> $\mathbf{v} = \mathbf{u} + \mathbf{a}t$
> $= (-3\mathbf{i} + \mathbf{j}) + 3 \times (2\mathbf{i} + 3\mathbf{j})$
> $= (-3\mathbf{i} + \mathbf{j}) + (6\mathbf{i} + 9\mathbf{j})$
> $= (3\mathbf{i} + 10\mathbf{j})\,\text{m s}^{-1}$
> So the speed of $P = \sqrt{3^2 + 10^2} = \sqrt{109}$
> $= 10.4\,\text{m s}^{-1}$ to 3 s.f.

Using $\mathbf{v} = \mathbf{u} + \mathbf{a}t$

The speed of P is the magnitude of its velocity.

- A force applied to a particle has both magnitude and direction, so force is also a vector. The force causes the particle to accelerate:

  $\mathbf{F} = m\mathbf{a}$, where m is the mass of the particle.

Example 14

A constant force, $\mathbf{F}\,\text{N}$, acts on a particle of mass $2\,\text{kg}$ for 10 seconds. The particle is initially at rest, and 10 seconds later it has velocity $(10\mathbf{i} - 24\mathbf{j})\,\text{m s}^{-1}$. Find $\mathbf{F}$.

> $(10\mathbf{i} - 24\mathbf{j}) = 10\mathbf{a}$
> So $\mathbf{a} = (\mathbf{i} - 2.4\mathbf{j})\,\text{m s}^{-2}$
> So $\mathbf{F} = 2 \times (\mathbf{i} - 2.4\mathbf{j}) = (2\mathbf{i} - 4.8\mathbf{j})\,\text{N}$

Use $\mathbf{v} = \mathbf{u} + \mathbf{a}t$ to find the acceleration of the particle. $\mathbf{u} = 0$ as the particle is initially at rest.

Using $\mathbf{F} = m\mathbf{a}$

Exercise 3D SKILLS PROBLEM-SOLVING

1. A particle P is moving with constant velocity $\mathbf{v}$ m s^{-1}. Initially P is at the point with position vector $\mathbf{r}$. Find the position of P t seconds later if:
   a $\mathbf{r} = 3\mathbf{j}$, $\mathbf{v} = 2\mathbf{i}$ and $t = 4$
   b $\mathbf{r} = 2\mathbf{i} - \mathbf{j}$, $\mathbf{v} = -2\mathbf{j}$ and $t = 3$
   c $\mathbf{r} = \mathbf{i} + 4\mathbf{j}$, $\mathbf{v} = -3\mathbf{i} + 2\mathbf{j}$ and $t = 6$
   d $\mathbf{r} = -3\mathbf{i} + 2\mathbf{j}$, $\mathbf{v} = 2\mathbf{i} - 3\mathbf{j}$ and $t = 5$

2. A particle P moves with constant velocity $\mathbf{v}$. Initially P is at the point with position vector $\mathbf{a}$. t seconds later P is at the point with position vector $\mathbf{b}$. Find $\mathbf{v}$ when:
   a $\mathbf{a} = 2\mathbf{i} + 3\mathbf{j}$, $\mathbf{b} = 6\mathbf{i} + 13\mathbf{j}$, $t = 2$
   b $\mathbf{a} = 4\mathbf{i} + \mathbf{j}$, $\mathbf{b} = 9\mathbf{i} + 16\mathbf{j}$, $t = 5$
   c $\mathbf{a} = 3\mathbf{i} - 5\mathbf{j}$, $\mathbf{b} = 9\mathbf{i} + 7\mathbf{j}$, $t = 3$
   d $\mathbf{a} = -2\mathbf{i} + 7\mathbf{j}$, $\mathbf{b} = 4\mathbf{i} - 8\mathbf{j}$, $t = 3$
   e $\mathbf{a} = -4\mathbf{i} + \mathbf{j}$, $\mathbf{b} = -12\mathbf{i} - 19\mathbf{j}$, $t = 4$

3. A particle moving with speed v m s^{-1} in direction $\mathbf{d}$ has velocity vector $\mathbf{v}$. Find $\mathbf{v}$ for these:
   a $v = 10$, $\mathbf{d} = 3\mathbf{i} - 4\mathbf{j}$
   b $v = 15$, $\mathbf{d} = -4\mathbf{i} + 3\mathbf{j}$
   c $v = 7.5$, $\mathbf{d} = -6\mathbf{i} + 8\mathbf{j}$
   d $v = 5\sqrt{2}$, $\mathbf{d} = \mathbf{i} + \mathbf{j}$
   e $v = 2\sqrt{13}$, $\mathbf{d} = -2\mathbf{i} + 3\mathbf{j}$
   f $v = \sqrt{68}$, $\mathbf{d} = 3\mathbf{i} - 5\mathbf{j}$
   g $v = \sqrt{60}$, $\mathbf{d} = -4\mathbf{i} - 2\mathbf{j}$
   h $v = 15$, $\mathbf{d} = -\mathbf{i} + 2\mathbf{j}$

4. A particle P starts at the point with position vector $\mathbf{r}_0$. P moves with constant velocity $\mathbf{v}$ m s^{-1}. After t seconds, P is at the point with position vector $\mathbf{r}$.
   a Find $\mathbf{r}$ if $\mathbf{r}_0 = 2\mathbf{i}$, $\mathbf{v} = \mathbf{i} + 3\mathbf{j}$, and $t = 4$.
   b Find $\mathbf{r}$ if $\mathbf{r}_0 = 3\mathbf{i} - \mathbf{j}$, $\mathbf{v} = -2\mathbf{i} + \mathbf{j}$, and $t = 5$.
   c Find $\mathbf{r}_0$ if $\mathbf{r} = 4\mathbf{i} + 3\mathbf{j}$, $\mathbf{v} = 2\mathbf{i} - \mathbf{j}$, and $t = 3$.
   d Find $\mathbf{r}_0$ if $\mathbf{r} = -2\mathbf{i} + 5\mathbf{j}$, $\mathbf{v} = -2\mathbf{i} + 3\mathbf{j}$, and $t = 6$.
   e Find $\mathbf{v}$ if $\mathbf{r}_0 = 2\mathbf{i} + 2\mathbf{j}$, $\mathbf{r} = 8\mathbf{i} - 7\mathbf{j}$, and $t = 3$.
   f Find the speed of P if $\mathbf{r}_0 = 10\mathbf{i} - 5\mathbf{j}$, $\mathbf{r} = -2\mathbf{i} + 9\mathbf{j}$, and $t = 4$.
   g Find t if $\mathbf{r}_0 = 4\mathbf{i} + \mathbf{j}$, $\mathbf{r} = 12\mathbf{i} - 11\mathbf{j}$, and $\mathbf{v} = 2\mathbf{i} - 3\mathbf{j}$.
   h Find t if $\mathbf{r}_0 = -2\mathbf{i} + 3\mathbf{j}$, $\mathbf{r} = 6\mathbf{i} - 3\mathbf{j}$, and the speed of P is 4 m s^{-1}.

5. The initial velocity of a particle P moving with uniform acceleration $\mathbf{a}$ m s^{-2} is $\mathbf{u}$ m s^{-1}. Find the velocity and the speed of P after t seconds in these cases:
   a $\mathbf{u} = 5\mathbf{i}$, $\mathbf{a} = 3\mathbf{j}$, and $t = 4$
   b $\mathbf{u} = 3\mathbf{i} - 2\mathbf{j}$, $\mathbf{a} = \mathbf{i} - \mathbf{j}$, and $t = 3$
   c $\mathbf{a} = 2\mathbf{i} - 3\mathbf{j}$, $\mathbf{u} = -2\mathbf{i} + \mathbf{j}$, and $t = 2$
   d $t = 6$, $\mathbf{u} = 3\mathbf{i} - 2\mathbf{j}$, and $\mathbf{a} = -\mathbf{i}$
   e $\mathbf{a} = 2\mathbf{i} + \mathbf{j}$, $t = 5$, and $\mathbf{u} = -3\mathbf{i} + 4\mathbf{j}$

6. A constant force $\mathbf{F}$ N acts on a particle of mass 4 kg for 5 seconds. The particle was initially at rest, and after 5 seconds it has velocity $(6\mathbf{i} - 8\mathbf{j})$ m s^{-1}. Find $\mathbf{F}$.

7. A force $(2\mathbf{i} - \mathbf{j})$ N acts on a particle of mass 2 kg. If the initial velocity of the particle is $(\mathbf{i} + 3\mathbf{j})$ m s^{-1}, find how far it moves in the first 3 seconds.

8 At time $t = 0$, a particle P is at the point with position vector $4\mathbf{i}$, and moving with constant velocity $(\mathbf{i} + \mathbf{j})\,\text{m s}^{-1}$. A second particle Q is at the point with position vector $-3\mathbf{j}$ and moving with velocity $\mathbf{v}\,\text{m s}^{-1}$. After 8 seconds, the paths of P and Q meet. Find the speed of Q.

(P) 9 At 2 pm the coastguard spots a rowing dinghy 500 m due south of his observation point. The dinghy has constant velocity $(2\mathbf{i} + 3\mathbf{j})\,\text{m s}^{-1}$.

> **Hint** In questions 9 and 10 the unit vectors $\mathbf{i}$ and $\mathbf{j}$ are due east and due north respectively.

   a Find, in terms of t, the position vector of the dinghy t seconds after 2 pm.

   b Find the distance of the dinghy from the observation point at 2:05 pm.

(P) 10 At noon a ferry F is 400 m due north of an observation point O moving with constant velocity $(7\mathbf{i} + 7\mathbf{j})\,\text{m s}^{-1}$, and a speedboat S is 500 m due east of O, moving with constant velocity $(23\mathbf{i} + 15\mathbf{j})\,\text{m s}^{-1}$.

   a Write down the position vectors of F and S at time t seconds after noon.

   b Show that F and S will collide, and find the position vector of the point of collision.

(P) 11 At 8 am two ships A and B are at $\mathbf{r}_A = (\mathbf{i} + 3\mathbf{j})\,\text{km}$ and $\mathbf{r}_B = (5\mathbf{i} - 2\mathbf{j})\,\text{km}$ from a fixed point P. Their velocities are $\mathbf{v}_A = (2\mathbf{i} - \mathbf{j})\,\text{km h}^{-1}$ and $\mathbf{v}_B = (-\mathbf{i} + 4\mathbf{j})\,\text{km h}^{-1}$ respectively.

   a Write down the position vectors of A and B t hours later.

   b Show that t hours after 8 am the position vector of B relative to A is given by $((4 - 3t)\mathbf{i} + (-5 + 5t)\mathbf{j})\,\text{km}$.

   c Show that the two ships do not collide.

   d Find the distance between A and B at 10 am.

(P) 12 A particle A starts at the point with position vector $12\mathbf{i} + 12\mathbf{j}$. The initial velocity of A is $(-\mathbf{i} + \mathbf{j})\,\text{m s}^{-1}$, and it has constant acceleration $(2\mathbf{i} - 4\mathbf{j})\,\text{m s}^{-2}$. Another particle, B, has initial velocity $\mathbf{i}\,\text{m s}^{-1}$ and constant acceleration $2\mathbf{j}\,\text{m s}^{-2}$. After 3 seconds the two particles collide. Find:

   a the speeds of the two particles when they collide

   b the position vector of the point where the two particles collide

   c the position vector of B's starting point.

Challenge   SKILLS CREATIVITY

During an air show, a stunt aeroplane passes over a control tower with velocity $(20\mathbf{i} - 100\mathbf{j})\,\text{m s}^{-1}$, and flies in a horizontal plane with constant acceleration $6\mathbf{j}\,\text{m s}^{-2}$. A second aeroplane passes over the same control tower at time t seconds later, where $t > 0$, travelling with velocity $(70\mathbf{i} + 40\mathbf{j})\,\text{m s}^{-1}$. The second aeroplane is flying in a horizontal plane with constant acceleration $-8\mathbf{j}\,\text{m s}^{-2}$.

Given that the two aeroplanes pass directly over one another in their subsequent motion, find the value of t.

Chapter review 3   SKILLS   PROBLEM-SOLVING

1. A coastguard station O monitors the movements of ships in a channel. At noon, the station's radar records two ships moving with constant speed. Ship A is at the point with position vector $(-3\mathbf{i} + 10\mathbf{j})$ km relative to O and has velocity $(2\mathbf{i} + 2\mathbf{j})$ km h^{-1}. Ship B is at the point with position vector $(6\mathbf{i} + \mathbf{j})$ km and has velocity $(-\mathbf{i} + 5\mathbf{j})$ km h^{-1}.

   Hint   In this question, the horizontal unit vectors $\mathbf{i}$ and $\mathbf{j}$ are directed due east and due north respectively.

   a Show that if the two ships maintain these velocities they will collide.

   The coastguard radios ship A and orders it to reduce its speed to move with velocity $(\mathbf{i} + \mathbf{j})$ km h^{-1}.

   Given that A obeys this order and maintains this new constant velocity:

   b find an expression for the vector $\overrightarrow{AB}$ at time t hours after noon

   c find, to three significant figures, the distance between A and B at 1500 hours

   d find the time at which B will be due north of A.

2. Two ships P and Q are moving along straight lines with constant velocities. Initially P is at a point O and the position vector of Q relative to O is $(12\mathbf{i} + 6\mathbf{j})$ km, where $\mathbf{i}$ and $\mathbf{j}$ are unit vectors directed due east and due north respectively. Ship P is moving with velocity $6\mathbf{i}$ km h^{-1} and ship Q is moving with velocity $(-3\mathbf{i} + 6\mathbf{j})$ km h^{-1}. At time t hours the position vectors of P and Q relative to O are $\mathbf{p}$ km and $\mathbf{q}$ km respectively.

   a Find $\mathbf{p}$ and $\mathbf{q}$ in terms of t.

   b Calculate the distance of Q from P when $t = 4$.

   c Calculate the value of t when Q is due north of P.

3. A particle P moves with constant acceleration $(-3\mathbf{i} + \mathbf{j})$ m s^{-2}. At time t seconds, its velocity is $\mathbf{v}$ m s^{-1}. When $t = 0$, $\mathbf{v} = 5\mathbf{i} - 3\mathbf{j}$.

   a Find the value of t when P is moving parallel to the vector $\mathbf{i}$.

   b Find the speed of P when $t = 5$.

   c Find the angle between the vector $\mathbf{i}$ and the direction of motion of P when $t = 5$.

4. A particle P of mass 5 kg is moving under the action of a constant force $\mathbf{F}$ N. At $t = 0$, P has velocity $(5\mathbf{i} - 3\mathbf{j})$ m s^{-1}. At $t = 4$ s, the velocity of P is $(-11\mathbf{i} + 5\mathbf{j})$ m s^{-1}. Find:

   a the acceleration of P in terms of $\mathbf{i}$ and $\mathbf{j}$

   b the magnitude of $\mathbf{F}$.

   At $t = 6$ s, P is at the point A with position vector $(28\mathbf{i} + 6\mathbf{j})$ m relative to a fixed origin O. At this instant the force $\mathbf{F}$ N is removed and P then moves with constant velocity. Two seconds after the force has been removed, P is at the point B.

   c Calculate the distance of B from O.

5. Two boats A and B are moving with constant velocities. Boat A moves with velocity $6\mathbf{i}$ km h^{-1}. Boat B moves with velocity $(3\mathbf{i} + 5\mathbf{j})$ km h^{-1}.

   Hint   In this question the vectors $\mathbf{i}$ and $\mathbf{j}$ are horizontal unit vectors in the directions due east and due north respectively.

   a Find the bearing on which B is moving.

At noon, A is at point O and B is 10 km due south of O. At time t hours after noon, the position vectors of A and B relative to O are $\mathbf{a}$ km and $\mathbf{b}$ km respectively.

b Find expressions for $\mathbf{a}$ and $\mathbf{b}$ in terms of t, giving your answer in the form $p\mathbf{i} + q\mathbf{j}$.

c Find the time when A is due east of B.

At time t hours after noon, the distance between A and B is d km.

d By finding an expression for $\overrightarrow{AB}$, show that $d^2 = 34t^2 - 100t + 100$.

At noon, the boats are 10 km apart.

e Find the time after noon at which the boats are again 10 km apart.

6 A small boat S, drifting in the sea, is modelled as a particle moving in a straight line at constant speed. When first sighted at 09:00, S is at a point with position vector $(-2\mathbf{i} - 4\mathbf{j})$ km relative to a fixed origin O, where $\mathbf{i}$ and $\mathbf{j}$ are unit vectors due east and due north respectively.

At 09:40, S is at the point with position vector $(4\mathbf{i} - 6\mathbf{j})$ km. At time t hours after 09:00, S is at the point with position vector $\mathbf{s}$ km.

a Calculate the bearing on which S is drifting.

b Find an expression for $\mathbf{s}$ in terms of t.

At 11:00 a motor boat M leaves O and travels with constant velocity $(p\mathbf{i} + q\mathbf{j})$ km h^{-1}.

c Given that M intercepts S at 11:30, calculate the value of p and the value of q.

7 A particle P moves in a horizontal plane. The acceleration of P is $(-2\mathbf{i} + 3\mathbf{j})$ m s^{-2}. At time $t = 0$, the velocity of P is $(3\mathbf{i} - 2\mathbf{j})$ m s^{-1}.

a Find, to the nearest degree, the angle between the vector $\mathbf{j}$ and the direction of motion of P when $t = 0$.

At time t seconds, the velocity of P is $\mathbf{v}$ m s^{-1}. Find:

b an expression for $\mathbf{v}$ in terms of t, in the form $a\mathbf{i} + b\mathbf{j}$

c the speed of P when $t = 4$

d the time when P is moving parallel to $\mathbf{i} + \mathbf{j}$.

8 At time $t = 0$ a football player kicks a ball from the point A with position vector $(3\mathbf{i} + 2\mathbf{j})$ m on a horizontal football field. The motion of the ball is modelled as that of a particle moving horizontally with constant velocity $(4\mathbf{i} + 9\mathbf{j})$ m s^{-1}. Find:

Hint In this question, the unit vectors $\mathbf{i}$ and $\mathbf{j}$ are horizontal vectors due east and due north respectively.

a the speed of the ball

b the position vector of the ball after t seconds.

The point B on the field has position vector $(29\mathbf{i} + 12\mathbf{j})$ m.

c Find the time when the ball is due north of B.

At time $t = 0$, another player starts running due north from B and moves with constant speed v m s^{-1}.

d Given that he intercepts the ball, find the value of v.

9 Two ships P and Q are travelling at night with constant velocities. At midnight, P is at the point with position vector $(10\mathbf{i} + 15\mathbf{j})$ km relative to a fixed origin O. At the same time, Q is at the point with position vector $(-16\mathbf{i} + 26\mathbf{j})$ km. Three hours later, P is at the point with position vector $(25\mathbf{i} + 24\mathbf{j})$ km. The ship Q travels with velocity $12\mathbf{i}$ km h^{-1}. At time t hours after midnight, the position vectors of P and Q are $\mathbf{p}$ km and $\mathbf{q}$ km respectively. Find:

a the velocity of P in terms of $\mathbf{i}$ and $\mathbf{j}$

b expressions for $\mathbf{p}$ and $\mathbf{q}$ in terms of t, $\mathbf{i}$ and $\mathbf{j}$.

At time t hours after midnight, the distance between P and Q is d km.

c By finding an expression for $\overrightarrow{PQ}$, show that $d^2 = 58t^2 - 430t + 797$.

Weather conditions are such that an observer on P can see the lights on Q only when the distance between P and Q is 13 km or less.

d Given that when $t = 2$ the lights on Q move into sight of the observer, find the time, to the nearest minute, at which the lights on Q move out of sight of the observer.

10 The velocity of a car is given by $\mathbf{v} = (12\mathbf{i} - 10\mathbf{j})$ m s^{-1}. Find:

a the speed of the car

b the angle that the direction of motion of the car makes with the unit vector $\mathbf{i}$.

11 The acceleration of a motorbike is given by $\mathbf{a} = (3\mathbf{i} - 4\mathbf{j})$ m s^{-2}. Find:

a the magnitude of the acceleration

b the angle that the direction of the acceleration vector makes with the unit vector $\mathbf{j}$.

Problem-solving

Draw a sketch to help you find the direction. $\mathbf{j}$ acts in the positive y-direction, so the angle between $\mathbf{j}$ and the vector $3\mathbf{i} - 4\mathbf{j}$ will be obtuse.

Challenge

The point B lies on the line with equation $3y = 15 - 5x$.

Given that $|\overrightarrow{OB}| = \dfrac{\sqrt{34}}{2}$, find two possible expressions for $\overrightarrow{OB}$ in the form $p\mathbf{i} + q\mathbf{j}$.

Summary of key points

1. A vector is a quantity which has both magnitude and direction.

2. The unit vectors along the Cartesian axes are usually denoted by **i** and **j** respectively. You can write any two-dimensional vector in the form $a\mathbf{i} + b\mathbf{j}$.

3. When vectors are written in terms of the unit vectors **i** and **j** you can add them together by adding the terms in **i** and the terms in **j** separately. You subtract vectors in a similar way.

4. When a vector is given in terms of the unit vectors **i** and **j** you can find its magnitude using Pythagoras' Theorem. The magnitude of a vector **a** is written |a|.

5. The velocity of a particle is a vector in the direction of motion. The magnitude of the velocity is the speed of the particle. The velocity is usually denoted by **v**.

6. If a particle starts from the point with position vector $\mathbf{r}_0$ and moves with constant velocity **v**, its displacement from its initial position at time t is $\mathbf{v}t$ and its position vector **r** is given by

    $\mathbf{r} = \mathbf{r}_0 + \mathbf{v}t$

7. The acceleration of a particle tells you how the velocity changes with time. Acceleration is a vector, usually denoted by **a**. If a particle with initial velocity **u** moves with constant acceleration **a** then its velocity, **v**, at time t is given by

    $\mathbf{v} = \mathbf{u} + \mathbf{a}t$

8. A force applied to a particle has both a magnitude and a direction, so force is also a vector. The force causes the particle to accelerate:

    $\mathbf{F} = m\mathbf{a}$, where m is the mass of the particle

9. If a particle is resting in equilibrium then the resultant of all the forces acting on it is zero. This means that the sum of the vectors of the forces is the zero vector.

4 DYNAMICS OF A PARTICLE MOVING IN A STRAIGHT LINE

4.1
4.2

Learning objectives

After completing this chapter you should be able to:
- Draw force diagrams and calculate resultant forces → pages 55–57
- Understand and use Newton's first law of motion → pages 55–57
- Calculate resultant forces by adding vectors → pages 58–60
- Understand and use Newton's second law of motion, $F = ma$ → pages 60–64
- Apply Newton's second law to vector forces and acceleration → pages 64–67
- Solve problems involving connected particles → pages 67–70
- Understand and use Newton's third law of motion → pages 68–70
- Solve problems involving pulleys → pages 71-75

Prior knowledge check

1. Calculate:
   a $(2\mathbf{i} + \mathbf{j}) + (3\mathbf{i} - 4\mathbf{j})$
   b $(-\mathbf{i} + 3\mathbf{j}) - (3\mathbf{i} - \mathbf{j})$
   ← International GCSE Mathematics

2. The diagram shows a right-angled triangle.
   Work out:
   a the length of the hypotenuse
   b the size of the angle a.
   Give your answers correct to 1 d.p. ← International GCSE Mathematics

3. A car starts from rest and accelerates at a constant rate of 1.5 m s^{-2}.
   a Work out the velocity of the car after 12 seconds.

   After 12 seconds, the driver brakes, causing the car to decelerate at a constant rate of 1 m s^{-2}.
   b Calculate the distance the car travels from the instant the driver brakes until the car comes to rest. ← Mechanics 1, Chapter 2

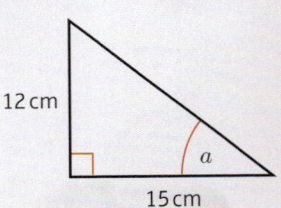

The weight of an air–sea rescue crew man is balanced by the tension in the cable. By modelling the forces in this situation, you can calculate how strong the cable needs to be.

DYNAMICS OF A PARTICLE MOVING IN A STRAIGHT LINE   CHAPTER 4

4.1 Force diagrams

A force diagram is a diagram showing all the forces acting on an object. Each force is shown as an arrow pointing in the direction in which the force acts. Force diagrams are used to model problems involving forces.

Example 1   SKILLS   INTERPRETATION

A block of weight W is being pulled to the right by a force, P, across a rough horizontal **plane**. Draw a force diagram to show all the forces acting on the block.

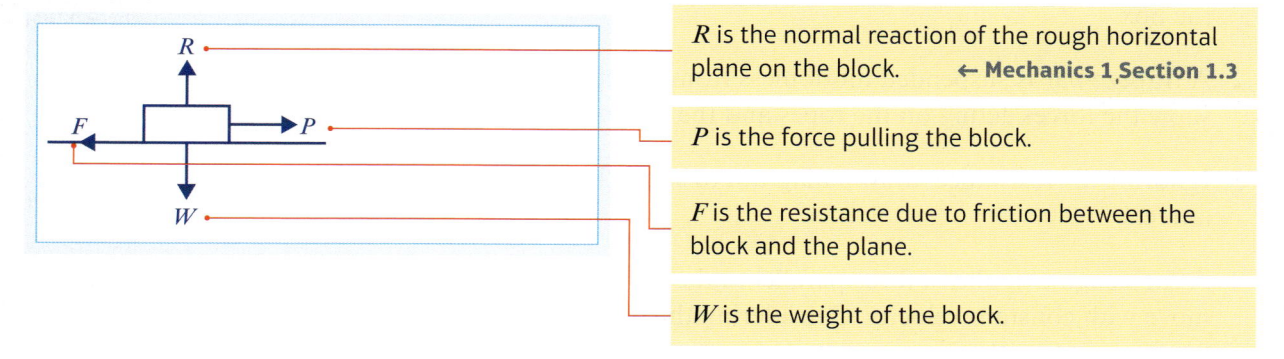

R is the normal reaction of the rough horizontal plane on the block.   ← Mechanics 1 Section 1.3

P is the force pulling the block.

F is the resistance due to friction between the block and the plane.

W is the weight of the block.

When the forces acting upon an object are balanced, the object is said to be in **equilibrium**.

- **Newton's first law of motion** states that an object at rest will stay at rest, and that an object moving with constant velocity will continue to move with constant velocity unless an unbalanced force acts on the object.

Watch out   Constant velocity means that neither the speed nor the direction is changing.

When there is more than one force acting on an object you can resolve the forces in a certain direction to find the resultant force in that direction. The direction you are resolving in becomes the positive direction. You add forces acting in this direction and subtract forces acting in the opposite direction.

In your answers, you can use the letter R, together with an arrow, R(↑), to indicate the direction in which you are resolving the forces.

In this section you will only resolve forces that are horizontal or vertical.

- **A resultant force acting on an object will cause the object to accelerate in the same direction as the resultant force.**

Example 2   SKILLS   INTERPRETATION

The diagram shows the forces acting on a particle.
a Draw a force diagram to represent the resultant force.
b Describe the motion of the particle.

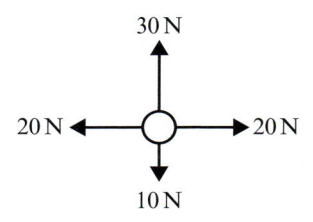

a 20 N

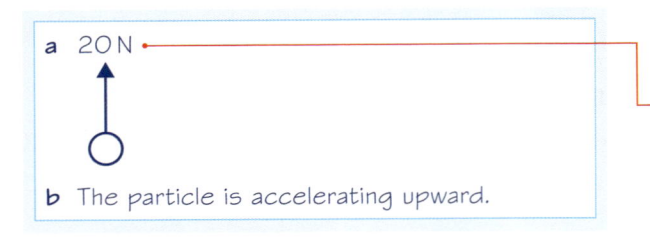

b The particle is accelerating upward.

R(→): 20 − 20 = 0, so the horizontal forces are balanced.
R(↑): 30 − 10 = 20, so the resultant force is 20 N upward.

Exercise 4A — SKILLS — INTERPRETATION

1. A box is at rest on a horizontal table. Draw a force diagram to show all the forces acting on the box.

2. A trapeze bar is suspended motionless from the ceiling by two ropes. Draw a force diagram to show the forces acting on the ropes and the trapeze bar.

3. Ignoring air resistance, draw a diagram to show the forces acting on an apple as it falls from a tree.

4. A car's engine applies a force parallel to the surface of a horizontal road that causes the car to move with constant velocity. Considering the resistance to motion, draw a diagram to show the forces acting on the car.

5. An air–sea rescue crew member is suspended motionless from a helicopter. Ignoring air resistance, show all the forces acting on him.

(P) 6. A satellite orbits the Earth at constant speed. State, with a reason, whether any resultant force is acting on the satellite.

Problem-solving
Consider the velocity of the satellite as it orbits the Earth.

7. A particle of weight 5 N sits at rest on a horizontal plane. State the value of the normal reaction acting on the particle.

8. Given that each of the particles is stationary, work out the value of P:

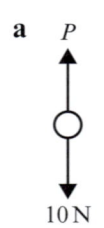

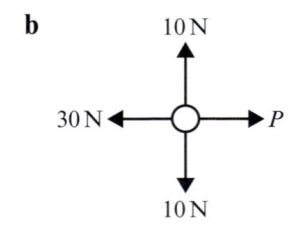

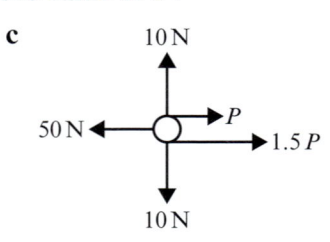

   a P up, 10 N down

   b 10 N up, 30 N left, P right, 10 N down

   c 10 N up, 50 N left, P right, $1.5P$ right, 10 N down

9. A platform is lifted vertically at constant velocity as shown in the diagram.

   a Ignoring air resistance, work out the tension, T in each rope.
   The tension in each rope is reduced by 50 N.

   b Describe the resulting motion of the platform.

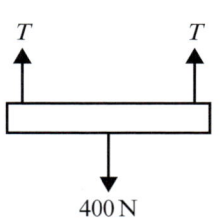

   T, T upward; 400 N downward

10 The diagram shows a particle acted on by a set of forces. Given that the particle is at rest, find the value of p and the value of q.

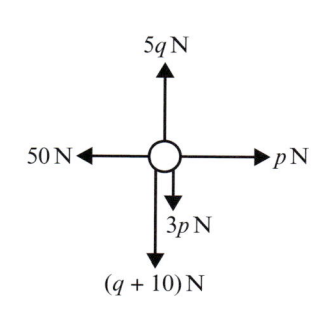

(P) 11 Given that the particle in this diagram is moving with constant velocity, v, find the values of P and Q.

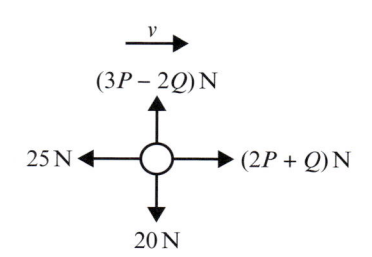

Problem-solving
Set up two simultaneous equations.

12 Each diagram below shows the forces acting on a particle.
   i Work out the size and direction of the resultant force.
   ii Describe the motion of the particle.

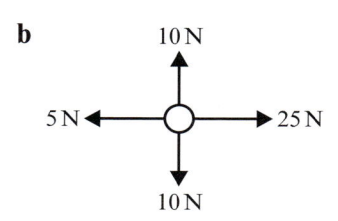

   a   b

13 A truck is moving along a horizontal level road. The truck's engine provides a forward thrust of 10 000 N. The total resistance is modelled as a constant force of magnitude 1600 N.
   a Modelling the truck as a particle, draw a force diagram to show the forces acting on the truck.
   b Calculate the resultant force acting on the truck.

(P) 14 A car is moving along a horizontal level road. The car's engine provides a constant driving force. The motion of the car is opposed by a constant resistance.
   a Modelling the car as a particle, draw a force diagram to show the forces acting on the car.
   b Given that the resultant force acting on the car is 4200 N in the direction of motion, and that the magnitude of the driving force is eight times the magnitude of the resistance force, calculate the magnitude of the resistance.

Problem-solving
Use algebra to describe the relationship between the driving force and the resistance.

4.2 Forces as vectors

You can write forces as vectors using **i–j** notation or as column vectors.

- **You can find the resultant of two or more forces given as vectors by adding the vectors.**

> **Links** If two forces $(p\mathbf{i} + q\mathbf{j})$ N and $(r\mathbf{i} + s\mathbf{j})$ N are acting on a particle, the resultant force will be $((p + r)\mathbf{i} + (q + s)\mathbf{j})$ N. ← Mechanics 1 Section 3.2

When a particle is in equilibrium the resultant vector force will be $0\mathbf{i} + 0\mathbf{j}$.

Example 3 — SKILLS: PROBLEM-SOLVING

The forces $2\mathbf{i} + 3\mathbf{j}$, $4\mathbf{i} - \mathbf{j}$, $-3\mathbf{i} + 2\mathbf{j}$ and $a\mathbf{i} + b\mathbf{j}$ act on an object which is in equilibrium. Find the values of a and b.

$(2\mathbf{i} + 3\mathbf{j}) + (4\mathbf{i} - \mathbf{j}) + (-3\mathbf{i} + 2\mathbf{j}) + (a\mathbf{i} + b\mathbf{j}) = 0$

$(2 + 4 - 3 + a)\mathbf{i} + (3 - 1 + 2 + b)\mathbf{j} = 0$

$\Rightarrow 3 + a = 0$ and $4 + b = 0$

$\Rightarrow a = -3$ and $b = -4$

If an object is in equilibrium then the resultant force will be zero.

*You can consider the **i** and **j** components separately because they are perpendicular.*

Example 4 — SKILLS: PROBLEM-SOLVING

In this question, **i** represents the unit vector due east, and **j** represents the unit vector due north. A particle begins at rest at the origin. It is acted on by three forces $(2\mathbf{i} + \mathbf{j})$ N, $(3\mathbf{i} - 2\mathbf{j})$ N and $(-\mathbf{i} + 4\mathbf{j})$ N.

a Find the resultant force in the form $p\mathbf{i} + q\mathbf{j}$.
b Work out the magnitude and bearing of the resultant force.
c Describe the motion of the particle.

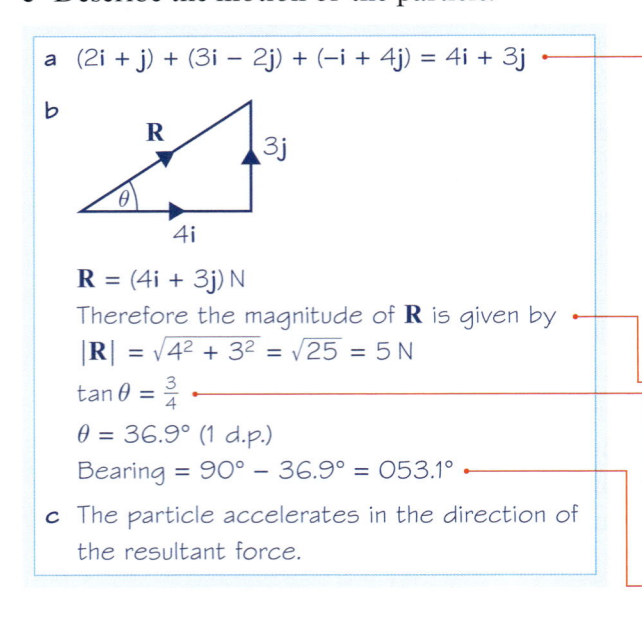

a $(2\mathbf{i} + \mathbf{j}) + (3\mathbf{i} - 2\mathbf{j}) + (-\mathbf{i} + 4\mathbf{j}) = 4\mathbf{i} + 3\mathbf{j}$

*Add together the **i**-components and the **j**-components.*

b

$\mathbf{R} = (4\mathbf{i} + 3\mathbf{j})$ N
Therefore the magnitude of **R** is given by
$|\mathbf{R}| = \sqrt{4^2 + 3^2} = \sqrt{25} = 5$ N
$\tan\theta = \frac{3}{4}$
$\theta = 36.9°$ (1 d.p.)
Bearing $= 90° - 36.9° = 053.1°$

c The particle accelerates in the direction of the resultant force.

*Notation: The unit vector **i** is usually taken to be due east or the positive x-direction. The unit vector **j** is usually taken to be due north or the positive y-direction. Questions involving finding bearings will often specify this.*

Use Pythagoras' theorem to find the magnitude of the resultant.

Use $\tan\theta = \dfrac{\text{opp}}{\text{adj}}$

Bearings are measured clockwise from north so subtract θ from $90°$.

DYNAMICS OF A PARTICLE MOVING IN A STRAIGHT LINE CHAPTER 4

Exercise 4B SKILLS PROBLEM-SOLVING

1 In each part of the question a particle is acted upon by the forces given. Work out the resultant force acting on the particle.

   a $(-\mathbf{i} + 3\mathbf{j})$ N and $(4\mathbf{i} - \mathbf{j})$ N

   b $\binom{5}{3}$ N and $\binom{-3}{-6}$ N

   c $(\mathbf{i} + \mathbf{j})$ N, $(5\mathbf{i} - 3\mathbf{j})$ N and $(-2\mathbf{i} - \mathbf{j})$ N

   d $\binom{-1}{4}$ N, $\binom{6}{0}$ N and $\binom{-2}{-7}$ N

> **Notation** $\binom{5}{3}$ N is the same as $(5\mathbf{i} + 3\mathbf{j})$ N.

2 An object is in equilibrium at O under the action of three forces $\mathbf{F}_1$, $\mathbf{F}_2$ and $\mathbf{F}_3$. Find $\mathbf{F}_3$ in these cases:

   a $\mathbf{F}_1 = (2\mathbf{i} + 7\mathbf{j})$ and $\mathbf{F}_2 = (-3\mathbf{i} + \mathbf{j})$

   b $\mathbf{F}_1 = (3\mathbf{i} - 4\mathbf{j})$ and $\mathbf{F}_2 = (2\mathbf{i} + 3\mathbf{j})$

(P) 3 The forces $\binom{a}{2b}$ N, $\binom{-2a}{-b}$ N and $\binom{3}{-4}$ N act on an object which is in equilibrium. Find the values of a and b.

4 For each force find:

   i the magnitude of the force

   ii the angle the force makes with **i**.

   a $(3\mathbf{i} + 4\mathbf{j})$ N   **b** $(5\mathbf{i} - \mathbf{j})$ N   **c** $(-2\mathbf{i} + 3\mathbf{j})$ N   **d** $\binom{-1}{-1}$ N

5 In this question, **i** represents the unit vector due east, and **j** represents the unit vector due north. A particle is acted upon by forces of:

   a $(-2\mathbf{i} + \mathbf{j})$ N, $(5\mathbf{i} + 2\mathbf{j})$ N and $(-\mathbf{i} - 4\mathbf{j})$ N

   b $(-2\mathbf{i} + \mathbf{j})$ N, $(2\mathbf{i} - 3\mathbf{j})$ N and $(3\mathbf{i} + 6\mathbf{j})$ N

Work out:

   i the resultant vector

   ii the magnitude of the resultant vector

   iii the bearing of the resultant vector.

(P) 6 The forces $(a\mathbf{i} - b\mathbf{j})$ N, $(b\mathbf{i} + a\mathbf{j})$ N and $(-4\mathbf{i} - 2\mathbf{j})$ N act on an object which is in equilibrium. Find the values of a and b.

> **Problem-solving** Use the **i** components and the **j** components to set up and solve two simultaneous equations.

(P) 7 The forces $(2a\mathbf{i} + 2b\mathbf{j})$ N, $(-5b\mathbf{i} + 3a\mathbf{j})$ N and $(-11\mathbf{i} - 7\mathbf{j})$ N act on an object which is in equilibrium. Find the values of a and b.

8 Three forces $\mathbf{F}_1$, $\mathbf{F}_2$ and $\mathbf{F}_3$ act on a particle. $\mathbf{F}_1 = (-3\mathbf{i} + 7\mathbf{j})$ N, $\mathbf{F}_2 = (\mathbf{i} - \mathbf{j})$ N and $\mathbf{F}_3 = (p\mathbf{i} + q\mathbf{j})$ N.

   a Given that this particle is in equilibrium, determine the value of p and the value of q.

The resultant of the forces $\mathbf{F}_1$ and $\mathbf{F}_2$ is $\mathbf{R}$.

   b Calculate, in N, the magnitude of $\mathbf{R}$.

   c Calculate, to the nearest degree, the angle between the line of action of $\mathbf{R}$ and the vector $\mathbf{j}$.

(E/P) 9 A particle is acted upon by two forces $\mathbf{F}_1$ and $\mathbf{F}_2$, given by $\mathbf{F}_1 = (3\mathbf{i} - 2\mathbf{j})$ N and $\mathbf{F}_2 = (a\mathbf{i} + 2a\mathbf{j})$ N, where a is a positive constant.

   a Find the angle between $\mathbf{F}_2$ and $\mathbf{i}$. **(2 marks)**

   The resultant of $\mathbf{F}_1$ and $\mathbf{F}_2$ is $\mathbf{R}$.

   b Given that $\mathbf{R}$ is parallel to $13\mathbf{i} + 10\mathbf{j}$, find the value of a. **(4 marks)**

(E) 10 Three forces $\mathbf{F}_1$, $\mathbf{F}_2$ and $\mathbf{F}_3$ acting on a particle P are given by the vectors $\mathbf{F}_1 = \begin{pmatrix} -7 \\ -4 \end{pmatrix}$ N, $\mathbf{F}_2 = \begin{pmatrix} 4 \\ 2 \end{pmatrix}$ N and $\mathbf{F}_3 = \begin{pmatrix} a \\ b \end{pmatrix}$ N, where a and b are constants.

   a Given that P is in equilibrium, find the value of a and the value of b. **(3 marks)**

   b The force $\mathbf{F}_1$ is now removed. The resultant of $\mathbf{F}_2$ and $\mathbf{F}_3$ is $\mathbf{R}$. Find:

     i the magnitude of $\mathbf{R}$ **(2 marks)**

     ii the angle, to the nearest degree, that the direction of $\mathbf{R}$ makes with the horizontal. **(3 marks)**

Challenge

An object is acted upon by a horizontal force of $10\mathbf{i}$ N and a vertical force $a\mathbf{j}$ N as shown in the diagram. The resultant of the two forces acts in the direction 60° to the horizontal. Work out the value of a and the magnitude of the resultant force.

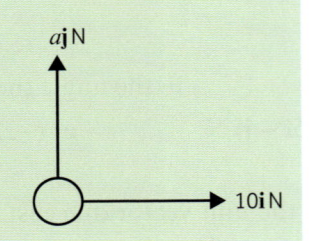

4.3 Forces and acceleration

A non-zero resultant force that acts on a particle will cause the particle to accelerate in the direction of the resultant force.

- **Newton's second law of motion** states that the force needed to accelerate a particle is equal to the product of the mass of the particle and the acceleration produced: $F = ma$.

A force of 1 N will accelerate a mass of 1 kg at a rate of 1 m s^{-2}. If a force F N acts on a particle of mass m kg causing it to accelerate at a m s^{-2}, the **equation of motion** for the particle is $F = ma$.

Gravity is the force between any object and the Earth. The force due to gravity acting on an object is called the **weight** of the object, and it acts vertically downward. A body falling freely experiences an acceleration of $g = 9.8$ m s^{-2}. Using the relationship $F = ma$ you can write the equation of motion for a body of mass m kg with weight W N.

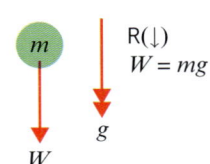

- $W = mg$

DYNAMICS OF A PARTICLE MOVING IN A STRAIGHT LINE   CHAPTER 4

Example 5   SKILLS   PROBLEM-SOLVING

Find the acceleration when a particle of mass 1.5 kg is acted on by a resultant force of 6 N.

$F = ma$
$6 = 1.5a$ — Substitute the values you know and solve the equation to find a.
$a = 4$
The acceleration is $4\,\text{m s}^{-2}$.

Example 6   SKILLS   PROBLEM-SOLVING

In each of these diagrams the body is accelerating as shown. Find the magnitudes of the unknown forces X and Y.

a Diagram: 2 kg block with $2\,\text{m s}^{-2}$ to the right; Y upward, $2g\,\text{N}$ downward, $4\,\text{N}$ left, X right.

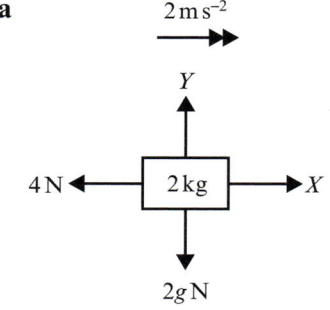

b Diagram: 4 kg block with $2\,\text{m s}^{-2}$ to the left; Y upward, $4g\,\text{N}$ downward, $20\,\text{N}$ downward, $80\,\text{N}$ left, X right.

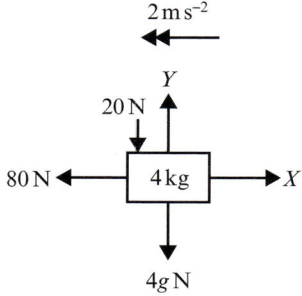

a $R(\rightarrow)$, $X - 4 = 2 \times 2$ — $R(\rightarrow)$ means that you are finding the resultant force in the horizontal direction, in the direction of the arrow. The arrow is in the positive direction.
$X = 8\,\text{N}$

$R(\uparrow)$, $Y - 2g = 2 \times 0$
$Y = 2g$
$\quad\; = 19.6\,\text{N}$ — This resultant force causes an acceleration of $2\,\text{m s}^{-2}$. Use $F = ma$.

b $R(\leftarrow)$, $80 - X = 4 \times 2$ — It is usually easier to take the positive direction as the direction of the acceleration.
$X = 72\,\text{N}$

$R(\uparrow)$, $Y - 20 - 4g = 4 \times 0$ — There is no vertical acceleration, so $a = 0$.
$Y = 20 + (4 \times 9.8)$
$\quad\; = 59.2\,\text{N}$ (3 s.f.)

Example 7   SKILLS   PROBLEM-SOLVING

A body of mass 5 kg is pulled along a rough horizontal table by a horizontal force of magnitude 20 N against a constant friction force of magnitude 4 N. Given that the body is initially at rest, find:
a the acceleration of the body
b the distance travelled by the body in the first 4 seconds
c the magnitude of the normal reaction between the body and the table.

a

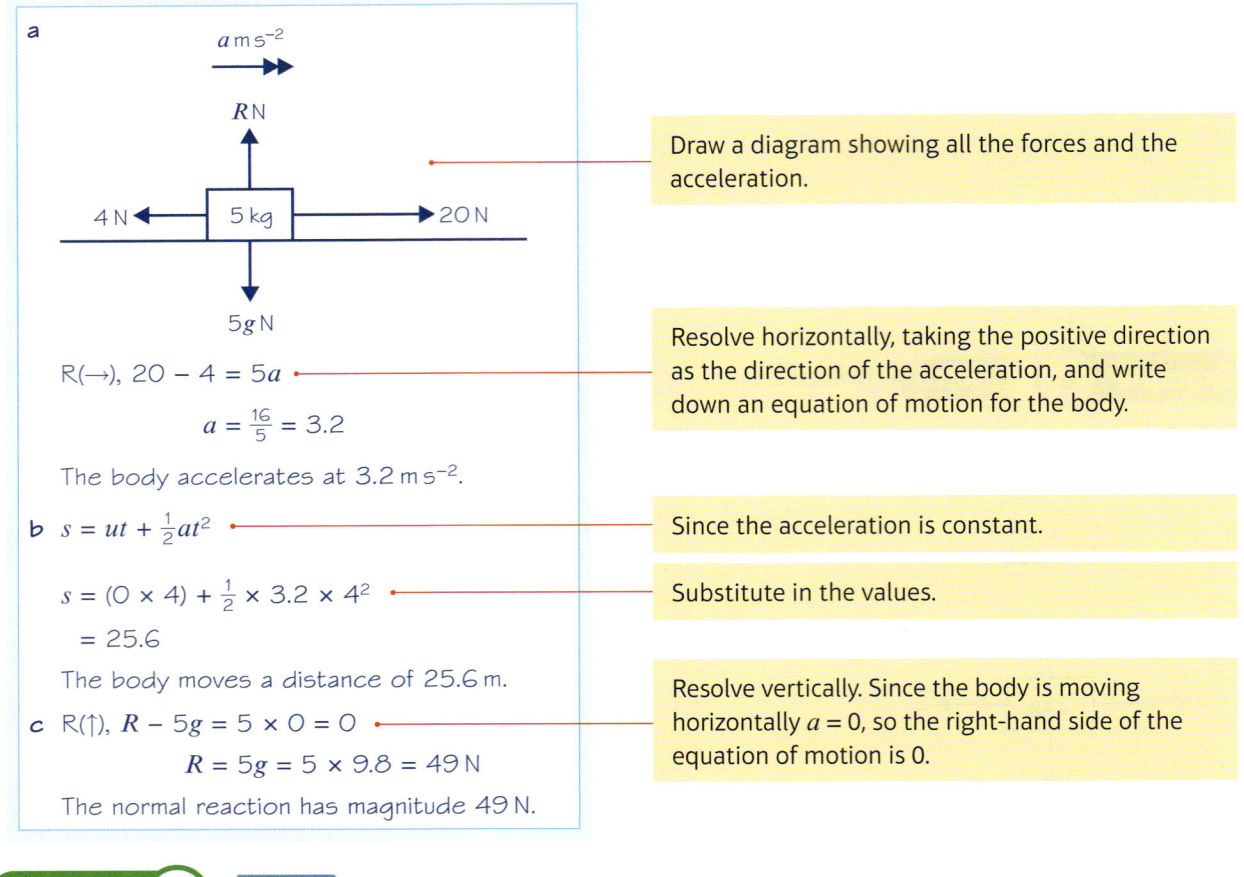

[Diagram: 5 kg block with acceleration a m s^{-2} to the right; forces: R N upward, 4 N left, 20 N right, $5g$ N downward]

Draw a diagram showing all the forces and the acceleration.

$R(\rightarrow)$, $20 - 4 = 5a$

Resolve horizontally, taking the positive direction as the direction of the acceleration, and write down an equation of motion for the body.

$a = \frac{16}{5} = 3.2$

The body accelerates at 3.2 m s^{-2}.

b $s = ut + \frac{1}{2}at^2$

Since the acceleration is constant.

$s = (0 \times 4) + \frac{1}{2} \times 3.2 \times 4^2$

Substitute in the values.

$= 25.6$

The body moves a distance of 25.6 m.

c $R(\uparrow)$, $R - 5g = 5 \times 0 = 0$

Resolve vertically. Since the body is moving horizontally $a = 0$, so the right-hand side of the equation of motion is 0.

$R = 5g = 5 \times 9.8 = 49$ N

The normal reaction has magnitude 49 N.

Exercise 4C — SKILLS — PROBLEM-SOLVING

1. Find the acceleration when a particle of mass 400 kg is acted on by a resultant force of 120 N.

2. Find the weight in newtons of a particle of mass 4 kg.

3. An object moving on a rough surface experiences a constant frictional force of 30 N which decelerates it at a rate of 1.2 m s^{-2}. Find the mass of the object.

(P) 4. An astronaut weighs 735 N on the Earth and 120 N on the moon. Work out the value of acceleration due to gravity on the moon.

Problem-solving

Start by finding the mass of the astronaut.

5. In each scenario, the forces acting on the body cause it to accelerate as shown. Find the magnitude of the unknown force.

a [2 kg block accelerating at 3 m s^{-2} upward; force P upward, $2g$ N downward]

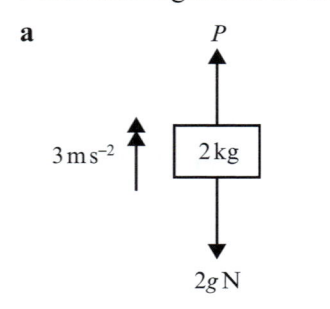

b [4 kg block accelerating at 2 m s^{-2} downward; force P upward, 10 N downward, $4g$ N downward]

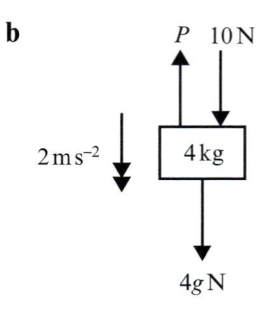

6 In each scenario, the forces acting on the body cause it to accelerate as shown. In each case find the mass of the body, m.

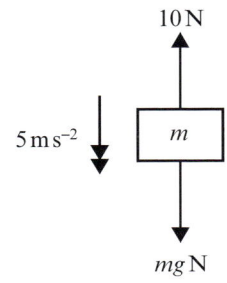

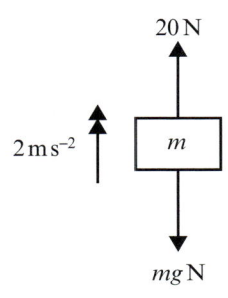

a 10 N ↑, $5\,\text{m s}^{-2}$ ↓, m, mg N ↓

b 20 N ↑, $2\,\text{m s}^{-2}$ ↑, m, mg N ↓

7 In each scenario, the forces acting on the body cause it to accelerate with magnitude $a\,\text{m s}^{-2}$. In each case find the value of a.

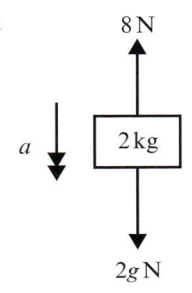

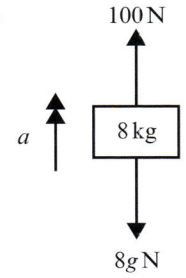

a 8 N ↑, a ↓, 2 kg, $2g$ N ↓

b 100 N ↑, a ↑, 8 kg, $8g$ N ↓

(P) 8 A horizontal force of 10 N acts upon a particle of mass 3 kg causing it to accelerate at $2\,\text{m s}^{-2}$ along a rough horizontal plane. Calculate the value of the force due to friction.

Problem-solving

Draw a force diagram showing all the forces acting on the particle.

(E/P) 9 A lift of mass 500 kg is lowered or raised by means of a metal cable attached to its top. The lift contains passengers whose total mass is 300 kg. The lift starts from rest and accelerates at a constant rate, reaching a speed of $3\,\text{m s}^{-1}$ after moving a distance of 5 m. Find:

  a the acceleration of the lift **(3 marks)**

Hint Use $v^2 = u^2 + 2as$.

  b the tension in the cable if the lift is moving vertically downward **(2 marks)**

  c the tension in the cable if the lift is moving vertically upward. **(2 marks)**

(E) 10 A trolley of mass 50 kg is pulled from rest in a straight line along a horizontal path by means of a horizontal rope attached to its front end. The trolley accelerates at a constant rate and after 2 s its speed is $1\,\text{m s}^{-1}$. As it moves, the trolley experiences a resistance to motion of magnitude 20 N. Find:

  a the acceleration of the trolley **(3 marks)**

  b the tension in the rope. **(2 marks)**

(E/P) 11 The engine of a van of mass 400 kg cuts out when it is moving along a straight horizontal road with speed $16\,\text{m s}^{-1}$. The van comes to rest without the brakes being applied.

In a model of the scenario it is assumed that the van is subject to a resistive force which has constant magnitude of 200 N.

a Find how long it takes the van to stop. **(3 marks)**
b Find how far the van travels before it stops. **(2 marks)**
c Comment on the suitability of the modelling assumption. **(1 mark)**

Challenge

A small stone of mass 400 g is projected vertically upward from the bottom of a pond full of water with speed 10 m s^{-1}. As the stone moves through the water it experiences a constant resistance of magnitude 3 N. Assuming that the stone does not reach the surface of the pond, find:
a the greatest height above the bottom of the pond that the stone reaches
b the speed of the stone as it hits the bottom of the pond on its return
c the total time taken for the stone to return to its initial position on the bottom of the pond.

4.4 Motion in two dimensions

- You can use **F** = m**a** to solve problems involving vector forces acting on particles.

> **Notation** In this version of the equation of motion, **F** and **a** are vectors.
> You can write acceleration as a 2D vector in the form $(p\mathbf{i} + q\mathbf{j})$ m s^{-2} or $\begin{pmatrix} p \\ q \end{pmatrix}$ m s^{-2}.

Example 8 SKILLS PROBLEM-SOLVING

In this question, **i** represents the unit vector due east, and **j** represents the unit vector due north.
A resultant force of $(3\mathbf{i} + 8\mathbf{j})$ N acts upon a particle of mass 0.5 kg.
a Find the acceleration of the particle in the form $(p\mathbf{i} + q\mathbf{j})$ m s^{-2}.
b Find the magnitude and bearing of the acceleration of the particle.

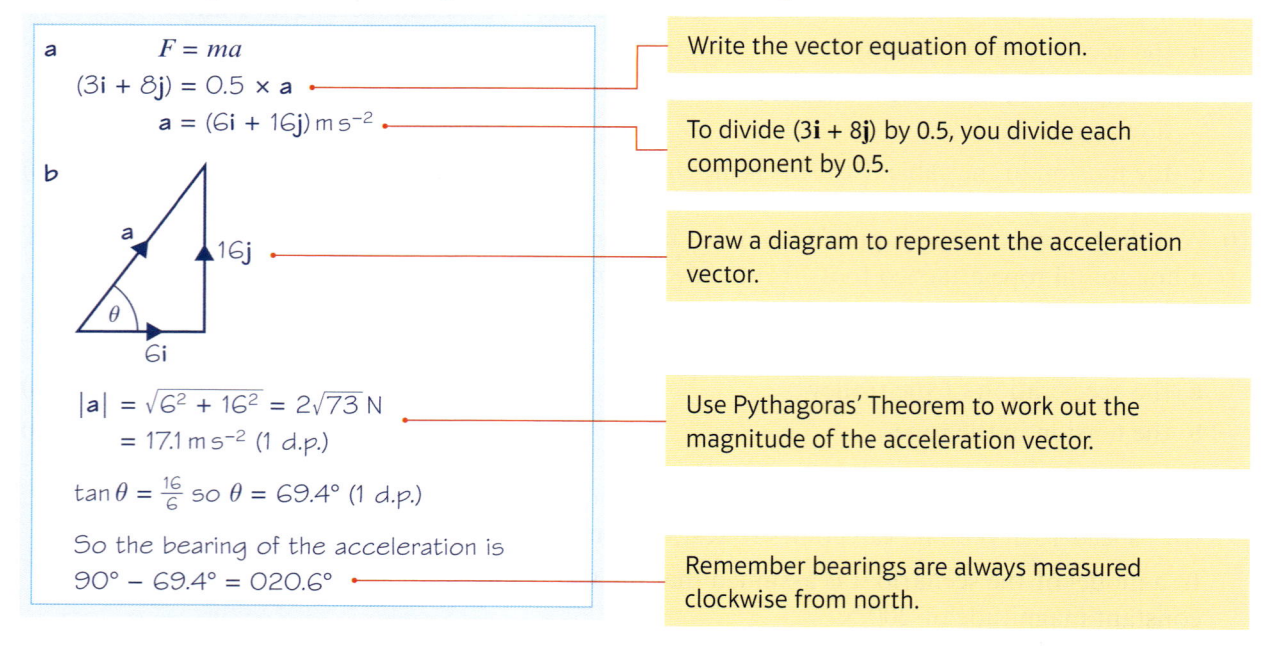

a $F = ma$ — Write the vector equation of motion.
$(3\mathbf{i} + 8\mathbf{j}) = 0.5 \times \mathbf{a}$
$\mathbf{a} = (6\mathbf{i} + 16\mathbf{j})$ m s^{-2} — To divide $(3\mathbf{i} + 8\mathbf{j})$ by 0.5, you divide each component by 0.5.

b

Draw a diagram to represent the acceleration vector.

$|\mathbf{a}| = \sqrt{6^2 + 16^2} = 2\sqrt{73}$ N
$= 17.1$ m s^{-2} (1 d.p.)

Use Pythagoras' Theorem to work out the magnitude of the acceleration vector.

$\tan \theta = \frac{16}{6}$ so $\theta = 69.4°$ (1 d.p.)

So the bearing of the acceleration is
$90° - 69.4° = 020.6°$

Remember bearings are always measured clockwise from north.

DYNAMICS OF A PARTICLE MOVING IN A STRAIGHT LINE — CHAPTER 4

Example 9 — SKILLS: PROBLEM-SOLVING

Forces $\mathbf{F}_1 = (2\mathbf{i} + 4\mathbf{j})$ N, $\mathbf{F}_2 = (-5\mathbf{i} + 4\mathbf{j})$ N, and $\mathbf{F}_3 = (6\mathbf{i} - 5\mathbf{j})$ N act on a particle of mass 3 kg. Find the acceleration of the particle.

Resultant force
$$= \mathbf{F}_1 + \mathbf{F}_2 + \mathbf{F}_3$$
$$= (2\mathbf{i} + 4\mathbf{j}) + (-5\mathbf{i} + 4\mathbf{j}) + (6\mathbf{i} - 5\mathbf{j})$$
$$= 3\mathbf{i} + 3\mathbf{j}$$
$3\mathbf{i} + 3\mathbf{j} = 3\mathbf{a} \Rightarrow \mathbf{a} = (\mathbf{i} + \mathbf{j})$ m s^{-2}

— Add the vectors to find the resultant force.

— Use $\mathbf{F} = m\mathbf{a}$.

Example 10 — SKILLS: PROBLEM-SOLVING

A boat is modelled as a particle of mass 60 kg being acted on by three forces:

$$\mathbf{F}_1 = \begin{pmatrix} 80 \\ 50 \end{pmatrix} \text{N} \qquad \mathbf{F}_2 = \begin{pmatrix} 10p \\ 20q \end{pmatrix} \text{N} \qquad \mathbf{F}_3 = \begin{pmatrix} -75 \\ 100 \end{pmatrix} \text{N}$$

Given that the boat is accelerating at a rate of $\begin{pmatrix} 0.8 \\ -1.5 \end{pmatrix}$ m s^{-2}, find the values of p and q.

Resultant force
$$= \mathbf{F}_1 + \mathbf{F}_2 + \mathbf{F}_3$$
$$= \begin{pmatrix} 80 \\ 50 \end{pmatrix} + \begin{pmatrix} 10p \\ 20q \end{pmatrix} + \begin{pmatrix} -75 \\ 100 \end{pmatrix}$$
$$= \begin{pmatrix} 5 + 10p \\ 150 + 20q \end{pmatrix} \text{N}$$

— Find the resultant force acting on the boat in terms of p and q.

$\mathbf{F} = m\mathbf{a}$
$$\begin{pmatrix} 5 + 10p \\ 150 + 20q \end{pmatrix} = 60 \times \begin{pmatrix} 0.8 \\ -1.5 \end{pmatrix} = \begin{pmatrix} 48 \\ -90 \end{pmatrix}$$

— Use $\mathbf{F} = m\mathbf{a}$. Remember that you need to multiply each component in the acceleration by 60.

So $5 + 10p = 48 \Rightarrow p = 4.3$
and $150 + 20q = -90 \Rightarrow q = -12$

— Solve separate equations for each component to find the values of p and q.

Exercise 4D — SKILLS: PROBLEM-SOLVING

In all the questions in this exercise, **i** represents the unit vector due east, and **j** represents the unit vector due north.

1. A resultant force of $(\mathbf{i} + 4\mathbf{j})$ N acts upon a particle of mass 2 kg.
   a. Find the acceleration of the particle in the form $(p\mathbf{i} + q\mathbf{j})$ m s^{-2}.
   b. Find the magnitude and bearing of the acceleration of the particle.

2. A resultant force of $(4\mathbf{i} + 3\mathbf{j})$ N acts on a particle of mass m kg causing it to accelerate at $(20\mathbf{i} + 15\mathbf{j})$ m s^{-2}. Work out the mass of the particle.

3 A particle of mass 3 kg is acted on by a force **F**. Given that the particle accelerates at $(7\mathbf{i} - 3\mathbf{j})$ m s^{-2}:

   a find an expression for **F** in the form $(p\mathbf{i} + q\mathbf{j})$ N

   b find the magnitude and bearing of **F**.

4 Two forces, $\mathbf{F}_1$ and $\mathbf{F}_2$, act on a particle of mass m. Find the acceleration of the particle, **a** m s^{-2}, given that:

   a $\mathbf{F}_1 = (2\mathbf{i} + 7\mathbf{j})$ N, $\mathbf{F}_2 = (-3\mathbf{i} + \mathbf{j})$ N, $m = 0.25$ kg

   b $\mathbf{F}_1 = (3\mathbf{i} - 4\mathbf{j})$ N, $\mathbf{F}_2 = (2\mathbf{i} + 3\mathbf{j})$ N, $m = 6$ kg

   c $\mathbf{F}_1 = (-40\mathbf{i} - 20\mathbf{j})$ N, $\mathbf{F}_2 = (25\mathbf{i} + 10\mathbf{j})$ N, $m = 15$ kg

   d $\mathbf{F}_1 = 4\mathbf{j}$ N, $\mathbf{F}_2 = (-2\mathbf{i} + 5\mathbf{j})$ N, $m = 1.5$ kg

   Notation You are asked to find the acceleration as a vector, **a**. You can give your answer as a column vector or using **i**–**j** notation.

5 A particle of mass 8 kg is at rest. It is acted on by three forces, $\mathbf{F}_1 = \begin{pmatrix} 3 \\ -1 \end{pmatrix}$ N, $\mathbf{F}_2 = \begin{pmatrix} 2 \\ -5 \end{pmatrix}$ N and $\mathbf{F}_3 = \begin{pmatrix} -1 \\ 0 \end{pmatrix}$ N.

   a Find the magnitude and direction of the acceleration of the particle, **a** m s^{-2}.

   b Find the time taken for the particle to travel a distance of 20 m.

   Hint Use $s = ut + \frac{1}{2}at^2$ with $s = 20$ and $u = 0$.

(E/P) 6 Two forces, $(2\mathbf{i} + 3\mathbf{j})$ N and $(p\mathbf{i} + q\mathbf{j})$ N, act on a particle P. The resultant of the two forces is **R**. Given that **R** acts in a direction which is parallel to the vector $(-\mathbf{i} + 4\mathbf{j})$, show that $4p + q + 11 = 0$. **(4 marks)**

   Problem-solving You can write **R** in the form $(-k\mathbf{i} + 4k\mathbf{j})$ N for some constant k.

(E) 7 A particle of mass 4 kg starts from rest and is acted upon by a force **R** of $(6\mathbf{i} + b\mathbf{j})$ N. **R** acts on a bearing of 045°.

   a Find the value of b. **(1 mark)**

   b Calculate the magnitude of **R**. **(2 marks)**

   c Work out the magnitude of the acceleration of the particle. **(2 marks)**

   d Find the total distance travelled by the particle during the first 5 seconds of its motion. **(3 marks)**

(P) 8 Three forces, $\mathbf{F}_1$, $\mathbf{F}_2$ and $\mathbf{F}_3$ act on a particle. $\mathbf{F}_1 = (-3\mathbf{i} + 7\mathbf{j})$ N, $\mathbf{F}_2 = (\mathbf{i} - \mathbf{j})$ N and $\mathbf{F}_3 = (p\mathbf{i} + q\mathbf{j})$ N.

   a Given that this particle is in equilibrium, determine the value of p and the value of q.

   Force $\mathbf{F}_2$ is removed.

   b Given that in the first 10 seconds of its motion the particle travels a distance of 12 m, find the exact mass of the particle in kg.

(P) 9 A particle of mass m kg is acted upon by forces of $(5\mathbf{i} + 6\mathbf{j})$ N, $(2\mathbf{i} - 2\mathbf{j})$ N and $(-\mathbf{i} - 4\mathbf{j})$ N, causing it to accelerate at 7 m s^{-2}. Work out the mass of the particle. Give your answer correct to 2 d.p.

E/P 10 Two forces, $\binom{2}{5}$ N and $\binom{p}{q}$ N, act on a particle P of mass m kg. The resultant of the two forces is **R**.

  a Given that **R** acts in a direction which is parallel to the vector $\binom{1}{-2}$, show that $2p + q + 9 = 0$. **(4 marks)**

  b Given also that $p = 1$ and that P moves with an acceleration of magnitude $15\sqrt{5}$ m s^{-2}, find the value of m. **(7 marks)**

> **Challenge**
>
> A particle of mass 0.5 kg is acted on by two forces:
> $\mathbf{F}_1 = -4\mathbf{i}$ N $\quad\quad \mathbf{F}_2 = (k\mathbf{i} + 2k\mathbf{j})$ N
> where k is a positive constant.
> Given that the particle is accelerating at a rate of $8\sqrt{17}$ m s^{-2}, find the value of k.

4.5 Connected particles

If a system involves the motion of more than one particle, the particles may be considered separately. However, if all parts of the system are moving in the **same straight line**, then you can also treat the whole system as a single particle.

- **You can solve problems involving connected particles by considering the particles separately or, if they are moving in the same straight line, as a single particle.**

> **Watch out** Particles need to remain in contact, or be connected by an inextensible rod or taut string to be considered as a single particle.

Example 11   SKILLS   PROBLEM-SOLVING

Two particles, P and Q, of masses 5 kg and 3 kg respectively, are connected by a light inextensible string. Particle P is pulled by a horizontal force of magnitude 40 N along a rough horizontal plane. Particle P experiences a frictional force of 10 N and particle Q experiences a frictional force of 6 N.

a Find the acceleration of the particles.
b Find the tension in the string.
c Explain how the modelling assumptions that the string is light and inextensible have been used.

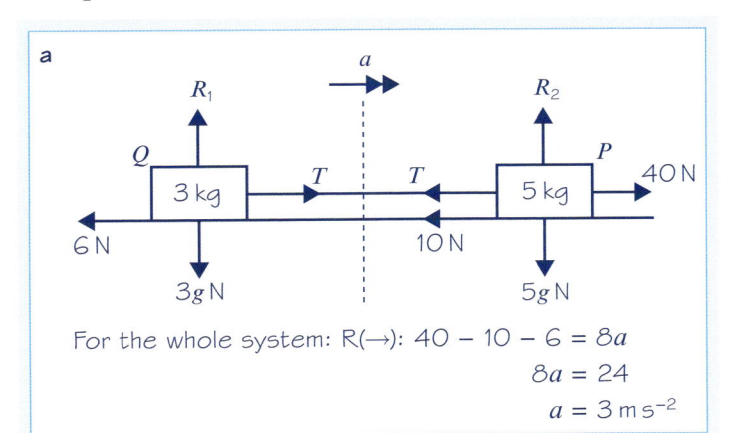

a

[Diagram: Q (3 kg) with R_1 up, $3g$ N down, 6 N left, T right; P (5 kg) with R_2 up, $5g$ N down, 10 N left, 40 N right, T left; acceleration a to the right]

For the whole system: R($\rightarrow$): $40 - 10 - 6 = 8a$
$8a = 24$
$a = 3$ m s^{-2}

> **Problem-solving**
>
> In part **a**, by considering the system as a single particle you eliminate the need to find the tension in the string. Otherwise you would need to set up two simultaneous equations involving a and T.
>
> In part **b**, the particles need to be considered separately to find the tension in the string.

b For P: R($\rightarrow$): $40 - T - 10 = 5 \times 3$

$T = 15$ N

c Inextensible $\Rightarrow$ acceleration of masses is the same.

light $\Rightarrow$ tension is the same throughout the length of the string and the mass of the string is negligible.

> You could also have chosen particle Q to find the tension. Check to see that it gives the same answer.

- **Newton's third law of motion** states that for every action there is an equal and opposite reaction.

Newton's third law means that when two bodies A and B are in contact, if body A exerts a force on body B, then body B exerts a force on body A that is equal in magnitude and acts in the opposite direction.

Example 12   SKILLS   PROBLEM-SOLVING

A light scale-pan is attached to a vertical light inextensible string. The scale-pan carries two masses A and B. The mass of A is 400 g and the mass of B is 600 g. A rests on top of B, as shown in the diagram.

The scale-pan is raised vertically, using the string, with acceleration 0.5 m s^{-2}.

a Find the tension in the string.

b Find the force exerted on mass B by mass A.

c Find the force exerted on mass B by the scale-pan.

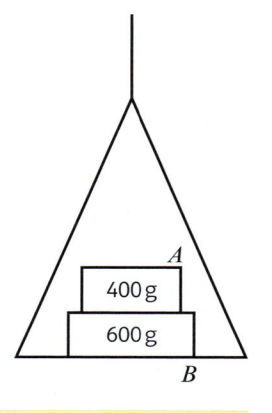

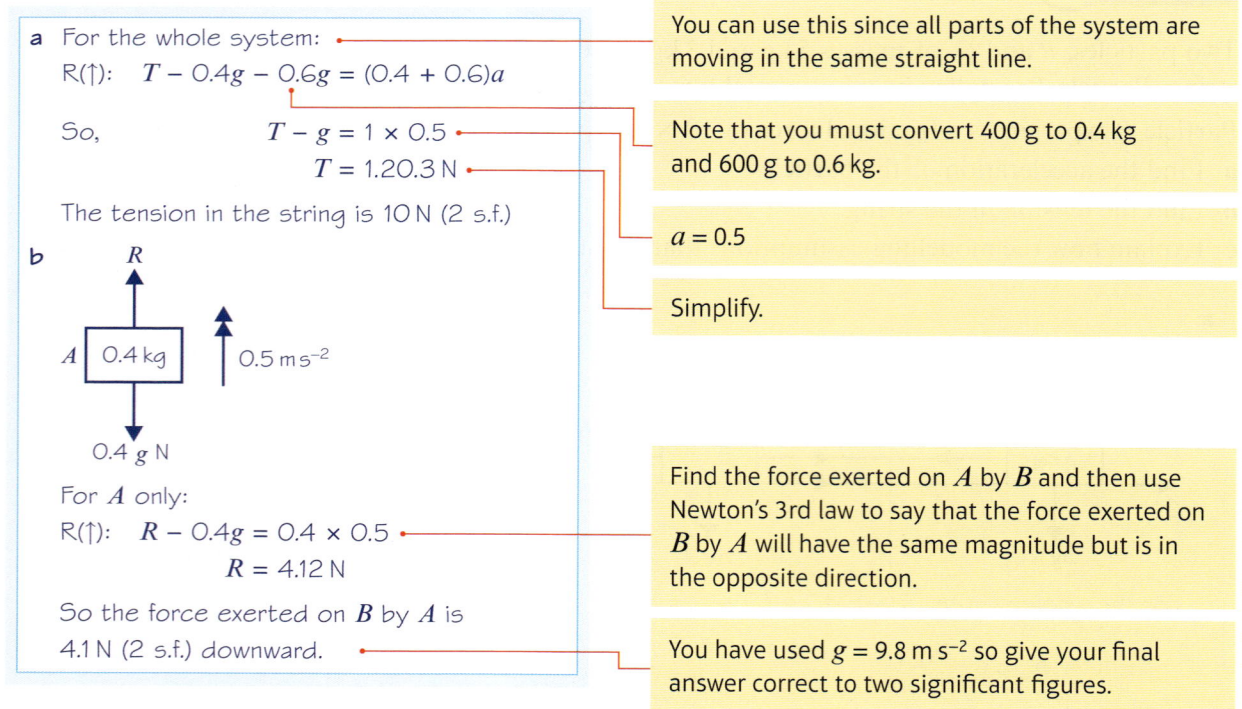

a For the whole system:

R($\uparrow$): $T - 0.4g - 0.6g = (0.4 + 0.6)a$

So, $T - g = 1 \times 0.5$

$T = 1.20.3$ N

The tension in the string is 10 N (2 s.f.)

> You can use this since all parts of the system are moving in the same straight line.

> Note that you must convert 400 g to 0.4 kg and 600 g to 0.6 kg.

> $a = 0.5$

> Simplify.

b

A 0.4 kg, R upward, 0.5 m s^{-2} upward, 0.4 g N downward

For A only:

R($\uparrow$): $R - 0.4g = 0.4 \times 0.5$

$R = 4.12$ N

So the force exerted on B by A is 4.1 N (2 s.f.) downward.

> Find the force exerted on A by B and then use Newton's 3rd law to say that the force exerted on B by A will have the same magnitude but is in the opposite direction.

> You have used $g = 9.8$ m s^{-2} so give your final answer correct to two significant figures.

c

For scale-pan only:

R(↑) $T - S = 0 \times 0.5$
           $= 0$

So, $T = S = 10.3$ N

So, the force exerted on *B* by the scale-pan is 10 N (2 s.f.) upward.

Problem-solving

It's easier to find the force exerted on the scale-pan by *B* and then use Newton's 3rd law to say that the force exerted on *B* by the scale-pan has the same magnitude but is in the opposite direction.

The scale-pan is light, i.e. its mass is 0.

From part **a**.

Use Newton's 3rd law.

Exercise 4E SKILLS PROBLEM-SOLVING

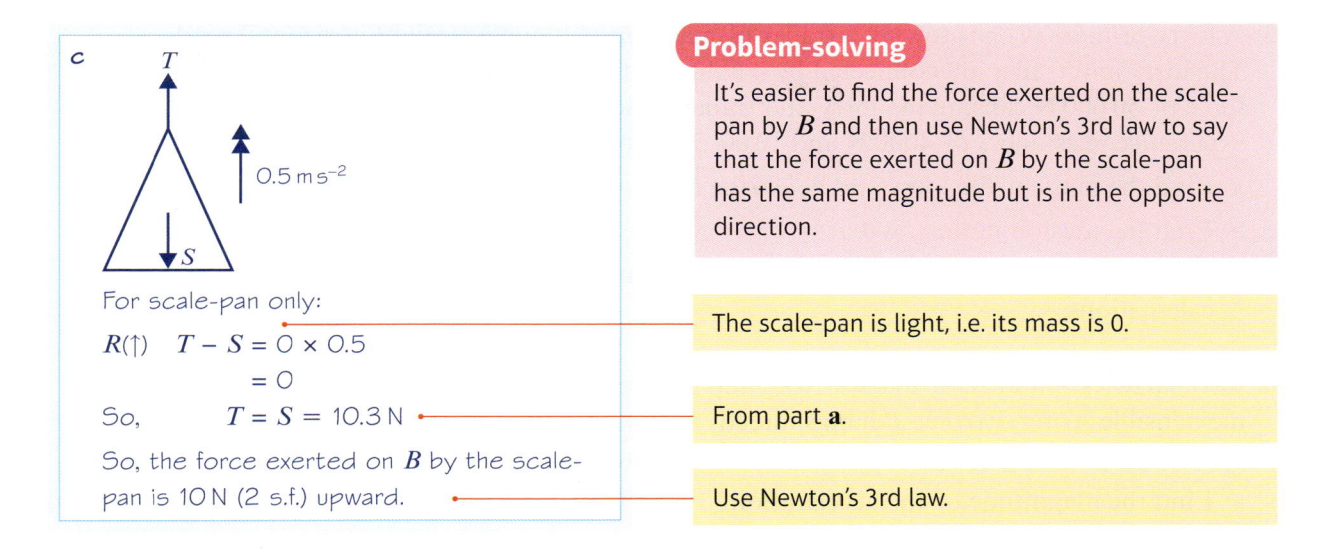

1 Two particles *P* and *Q*, of masses 8 kg and 2 kg respectively, are connected by a light inextensible string. The particles are on a smooth horizontal plane. A horizontal force of magnitude *F* is applied to *P* in a direction away from *Q* and when the string is taut the particles move with acceleration 0.4 m s^{-2}.

   a Find the value of *F*.

   b Find the tension in the string.

   c Explain how the modelling assumptions that the string is light and inextensible are used.

2 Two particles *P* and *Q* of masses 20 kg and *m* kg respectively, are connected by a light inextensible rod. The particles lie on a smooth horizontal plane. A horizontal force of 60 N is applied to *Q* in a direction toward *P*, causing the particles to move with acceleration 2 m s^{-2}.

   a Find the mass, *m*, of *Q*.

   b Find the thrust in the rod.

   Hint For part **b**, consider *P* on its own.

3 Two particles *P* and *Q*, of masses 7 kg and 8 kg respectively, are connected by a light inextensible string. The particles are on a smooth horizontal plane. A horizontal force of 30 N is applied to *Q* in a direction away from *P*. When the string is taut the particles move with acceleration $a \text{ m s}^{-2}$.

   a Find the acceleration, *a*, of the system.

   b Find the tension in the string.

4 Two boxes *A* and *B*, of masses 110 kg and 190 kg respectively, sit on the floor of a lift of mass 1700 kg. Box *A* rests on top of box *B*. The lift is supported by a light inextensible cable and is descending with constant acceleration 1.8 m s^{-2}:

   a Find the tension in the cable.

   b Find the force exerted by box *B*

   i on box *A*   **ii** on the floor of the lift.

5 A lorry of mass m kg is towing a trailer of mass $3m$ kg along a straight horizontal road. The lorry and trailer are connected by a light inextensible tow-bar. The lorry exerts a driving force of 50 000 N causing the lorry and trailer to accelerate at 5 m s^{-2}. The lorry and trailer experience resistances of 4000 N and 10 000 N respectively.

   a Find the mass of the lorry and hence the mass of the trailer.

   b Find the tension in the tow-bar.

   c Explain how the modelling assumptions that the tow-bar is light and inextensible affect your calculations.

6 Two particles A and B, of masses 10 kg and 5 kg respectively, are connected by a light inextensible string. Particle B hangs directly below particle A. A force of 180 N is applied to A vertically upward causing the particles to accelerate.

   a Find the magnitude of the acceleration. **(3 marks)**

   b Find the tension in the string. **(2 marks)**

7 Two particles A and B, of masses 6 kg and m kg respectively, are connected by a light inextensible string. Particle B hangs directly below particle A. A force of 118 N is applied to A vertically upward causing the particles to accelerate at 2 m s^{-2}.

   a Find the mass, m, of particle B. **(3 marks)**

   b Find the tension in the string. **(2 marks)**

8 A train engine of mass 6400 kg is pulling a carriage of mass 1600 kg along a straight horizontal railway track. The engine is connected to the carriage by a shunt which is parallel to the direction of motion of the coupling. The shunt is modelled as a light rod. The engine provides a constant driving force of 12 000 N. The resistances to the motion of the engine and the carriage are modelled as constant forces of magnitude R N and 2000 N respectively.

   Given that the acceleration of the engine and the carriage is 0.5 m s^{-2}:

   a find the value of R **(3 marks)**

   b show that the tension in the shunt is 2800 N. **(2 marks)**

9 A car of mass 900 kg pulls a trailer of mass 300 kg along a straight horizontal road using a light tow-bar which is parallel to the road. The horizontal resistances to motion of the car and the trailer have magnitudes 200 N and 100 N respectively. The engine of the car produces a constant horizontal driving force on the car of magnitude 1200 N.

   a Show that the acceleration of the car and trailer is 0.75 m s^{-2}. **(2 marks)**

   b Find the magnitude of the tension in the tow-bar. **(3 marks)**

   The car is moving along the road when the driver sees a set of traffic lights have turned red. He reduces the force produced by the engine to zero and applies the brakes. The brakes produce a force on the car of magnitude F newtons and the car and trailer decelerate.

   c Given that the resistances to motion are unchanged and the magnitude of the thrust in the tow-bar is 100 N, find the value of F. **(7 marks)**

4.6 Pulleys

In this section you will see how to model systems of connected particles involving pulleys. The problems you answer will assume that particles are connected by a light, inextensible string, which passes over a **smooth pulley**. This means that the tension in the string will be the same **on both sides** of the pulley. The parts of the string on each side of the pulley will be either horizontal or vertical. More complicated systems will be considered in Chapter 5.

Watch out You cannot treat a system involving a pulley as a single particle. This is because the particles are moving in different directions.

Example 13   SKILLS   PROBLEM-SOLVING

Particles P and Q, of masses $2m$ and $3m$, are attached to the ends of a light inextensible string. The string passes over a small smooth fixed pulley and the masses hang with the string taut. The system is released from rest.

a **i** Write down an equation of motion for P.
  ii Write down an equation of motion for Q.
b Find the acceleration of each mass.
c Find the tension in the string.
d Find the force exerted on the pulley by the string.
e Find the distance moved by Q in the first 4 s, assuming that P does not reach the pulley.
f State how you have used the fact that the pulley is smooth in your calculations.

a

[Diagram showing pulley with force F upward on pulley from ceiling, tensions T on both sides, particle P (mass $2m$) on left with weight $2mg$ N and acceleration a upward, particle Q (mass $3m$) on right with weight $3mg$ N and acceleration a downward.]

Problem-solving
Resolve vertically for both P and Q. This will give you simultaneous equations involving the tension T and the acceleration a which can then be solved.

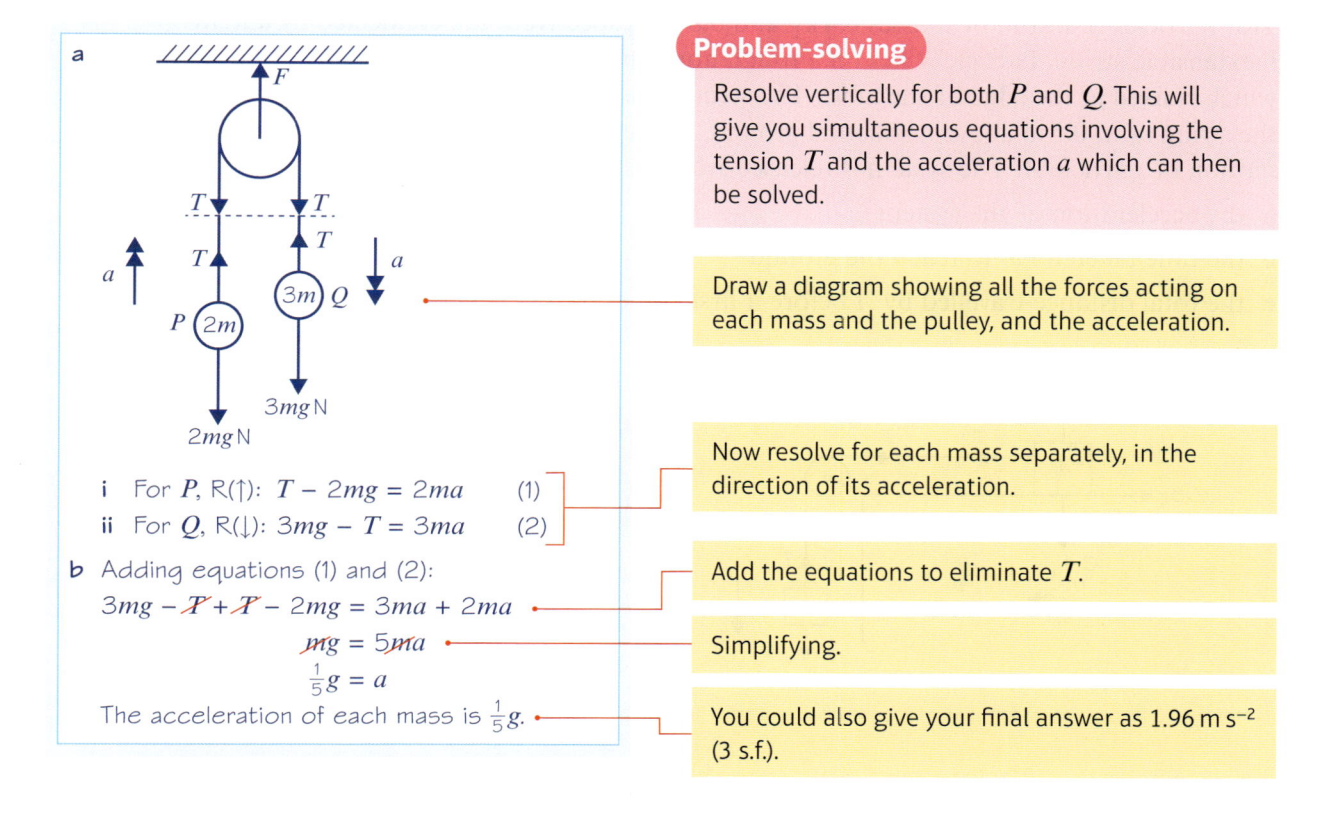

Draw a diagram showing all the forces acting on each mass and the pulley, and the acceleration.

  i For P, R($\uparrow$): $T - 2mg = 2ma$ (1)
  ii For Q, R($\downarrow$): $3mg - T = 3ma$ (2)

Now resolve for each mass separately, in the direction of its acceleration.

b Adding equations (1) and (2):
$$3mg - \cancel{T} + \cancel{T} - 2mg = 3ma + 2ma$$

Add the equations to eliminate T.

$$\cancel{m}g = 5\cancel{m}a$$

Simplifying.

$$\tfrac{1}{5}g = a$$

The acceleration of each mass is $\tfrac{1}{5}g$.

You could also give your final answer as 1.96 m s^{-2} (3 s.f.).

c From (1): $T - 2mg = 2m \times \frac{1}{5}g$ ← Substitute for a.

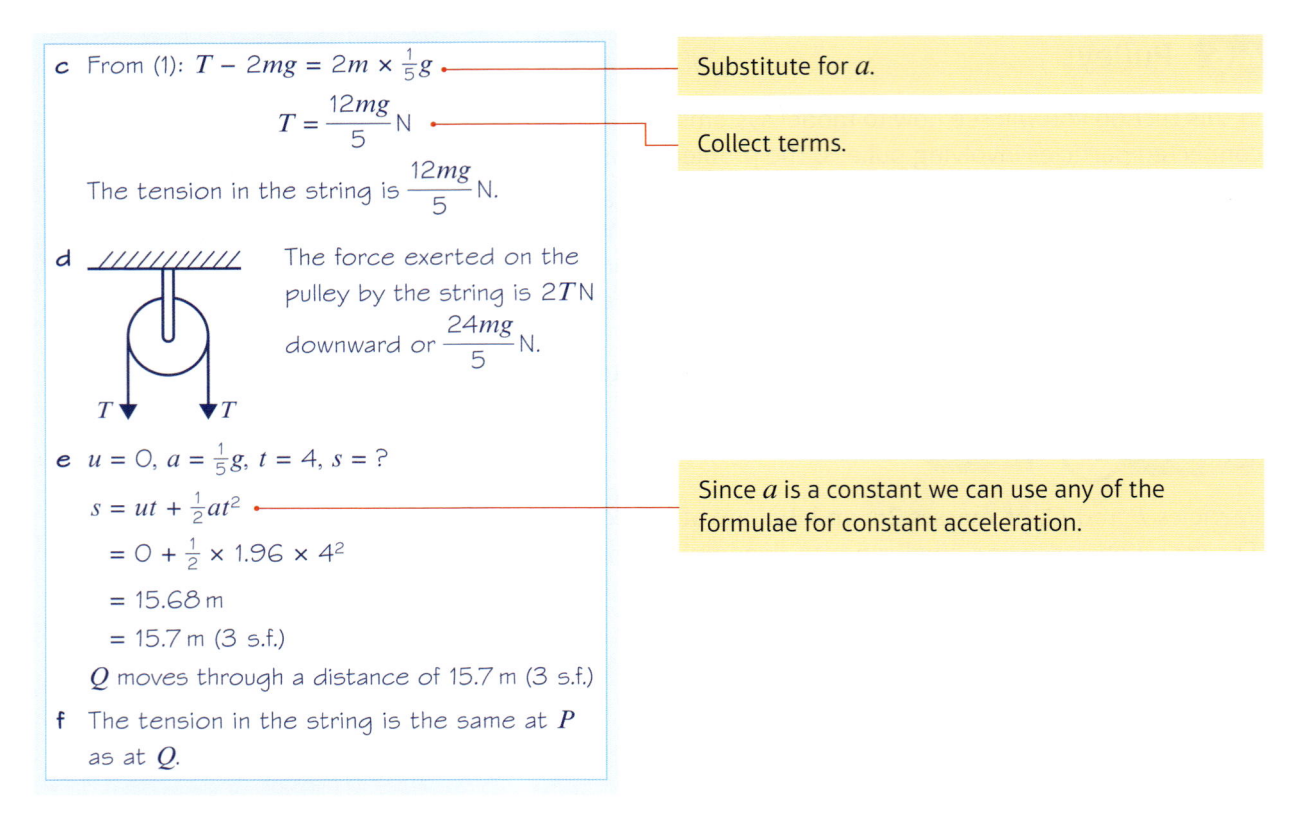

$$T = \frac{12mg}{5} \text{ N}$$ ← Collect terms.

The tension in the string is $\frac{12mg}{5}$ N.

d The force exerted on the pulley by the string is $2T$ N downward or $\frac{24mg}{5}$ N.

e $u = 0$, $a = \frac{1}{5}g$, $t = 4$, $s = ?$

$s = ut + \frac{1}{2}at^2$ ← Since a is a constant we can use any of the formulae for constant acceleration.

$= 0 + \frac{1}{2} \times 1.96 \times 4^2$

$= 15.68$ m

$= 15.7$ m (3 s.f.)

Q moves through a distance of 15.7 m (3 s.f.)

f The tension in the string is the same at P as at Q.

Example 14 — SKILLS: PROBLEM-SOLVING

Two particles A and B, of masses 0.4 kg and 0.8 kg respectively, are connected by a light inextensible string. Particle A lies on a rough horizontal table 4.5 m from a small smooth pulley which is fixed at the edge of the table. The string passes over the pulley and B hangs freely, with the string taut, 0.5 m above horizontal ground. A frictional force of magnitude $0.08g$ opposes the motion of particle A. The system is released from rest. Find:

a the acceleration of the system

b the time taken for B to reach the ground

c the total distance travelled by A before it first comes to rest.

a

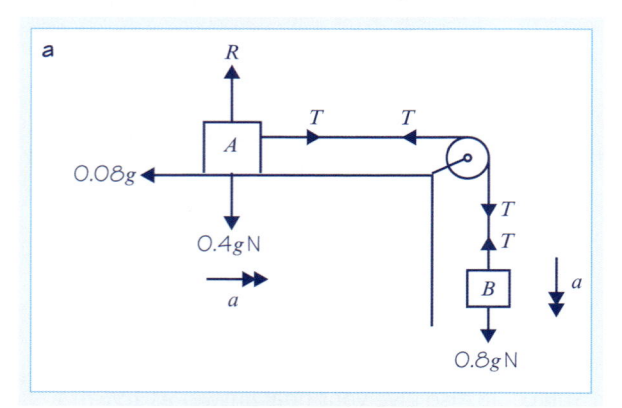

Problem-solving

Draw a diagram showing all the forces and the accelerations. The pulley is smooth so the tension in the string is the same on each side of the pulley.

For A only: $R(\rightarrow)$, $T - 0.08g = 0.4a$ (1) — Write equations of motion for each of A and B separately.

For B only: $R(\downarrow)$, $0.8g - T = 0.8a$ (2)

Add (1) and (2):

$0.8g - \cancel{T} + \cancel{T} - 0.08g = 0.8a + 0.4a$ — Eliminate the T terms.

$0.72g = 1.2a$

$0.6g = a$

The acceleration of the system is $0.6g$. — You could also give your answer as 5.88 m s^{-2} (3 s.f.).

b

$u = 0$, $s = 0.5$,

$a = 5.88$, $t = ?$ — Use an unrounded value of the acceleration.

$s = ut + \frac{1}{2}at^2$ — The acceleration is constant.

$0.5 = 0 + \frac{1}{2} \times 5.88 \times t^2$

$t = 0.412$ (3 s.f.)

The time taken for B to hit the ground is 0.412 s (3 s.f.)

c Find the speed of B when it hits the ground. — Use an unrounded value for t.

$u = 0$, $a = 5.88$, $t = 0.412\,39$, $v = ?$

$v = u + at$

$v_B = 0 + 5.88 \times 0.412\,39 = 2.424\,87$ m s^{-1}. — Using surds, $v_B = \sqrt{\dfrac{3g}{5}}$

Speed of A on the table is $2.424\,87$ m s^{-1}.

Once B hits the ground the string will go slack and A will begin to decelerate as it slides against the friction on the table. — Since the string is inextensible.

From (1): $-0.08g = 0.4a'$ — Put $T = 0$ in equation (1) as the string is now slack.

$a' = -0.2g$

$u_A = 2.424\,87$, $v = 0$, $a' = -0.2g$, $s = ?$ — This is the new acceleration of A along the table.

$v^2 = u^2 + 2as$

$0^2 = 2.424\,87^2 - 0.4gs$

$s = 1.50$ m (3 s.f.)

A slides a further 1.50 m along the table before it comes to rest.

∴ Total distance moved by A is

$0.5 + 1.50 = 2.00$ m (3 s.f.)

Exercise 4F   SKILLS   PROBLEM-SOLVING

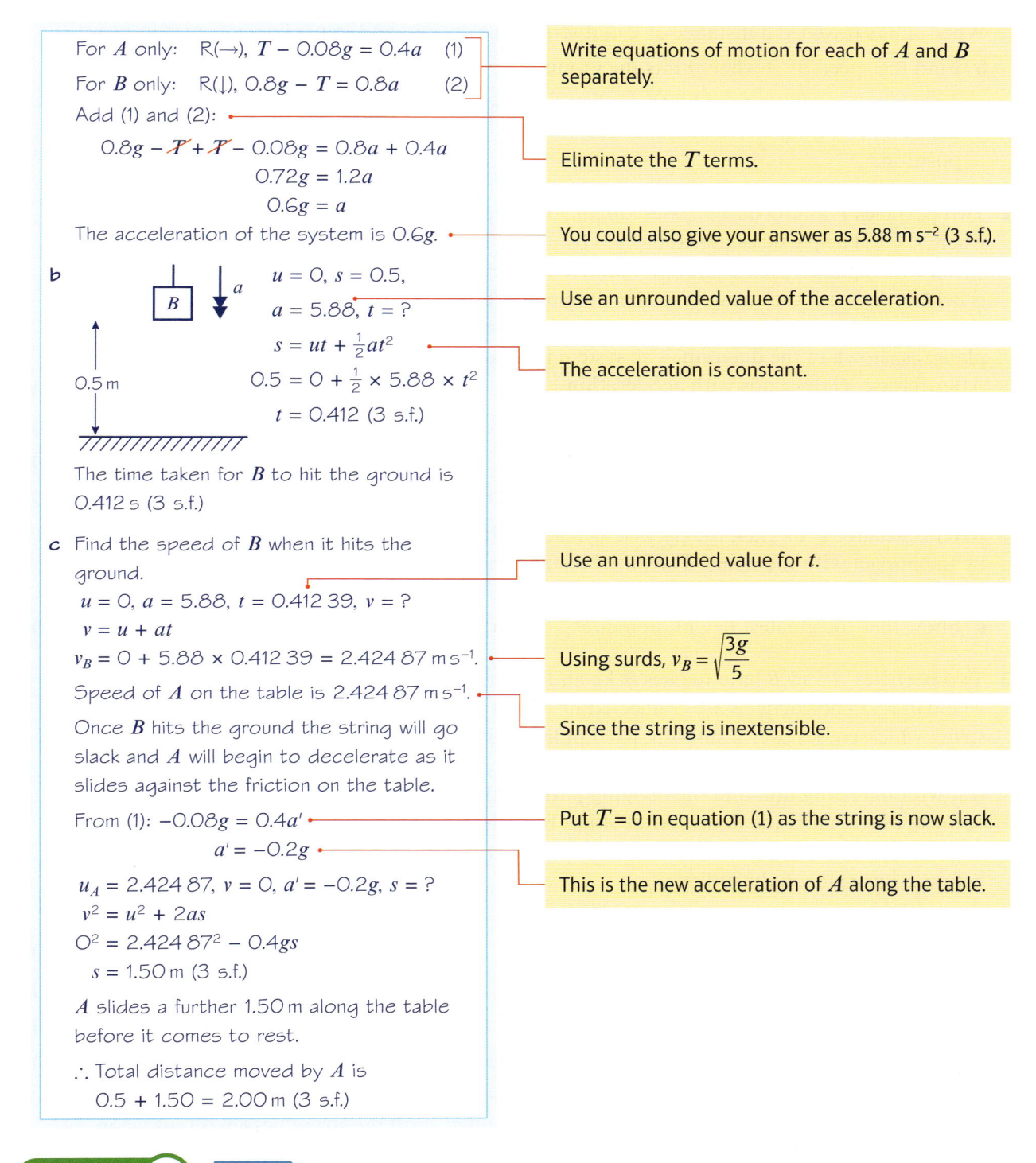

P 1 Two particles A and B, of masses 4 kg and 3 kg respectively, are connected by a light inextensible string which passes over a small smooth fixed pulley. The particles are released from rest with the string taut.

a Find the tension in the string.

When A has travelled a distance of 2 m it strikes the ground and immediately comes to rest.

b Find the speed of A when it hits the ground.

c Assuming that B does not hit the pulley, find the greatest height that B reaches above its initial position.

> **Problem-solving**
>
> After A hits the ground, B behaves like a particle moving freely under gravity.

E/P **2** Two particles P and Q have masses km and $3m$ respectively, where $k < 3$. The particles are connected by a light inextensible string which passes over a smooth light fixed pulley. The system is held at rest with the string taut, the hanging parts of the string vertical and with P and Q at the same height above a horizontal plane, as shown in the diagram. The system is released from rest. After release, Q descends with acceleration $\frac{1}{3}g$.

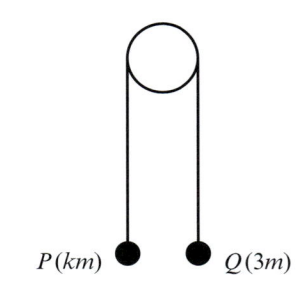

a Calculate the tension in the string as Q descends. **(3 marks)**

b Show that $k = 1.5$. **(3 marks)**

c State how you have used the information that the pulley is smooth. **(1 mark)**

After descending for 1.8 s, the particle Q reaches the plane. It is immediately brought to rest by the impact with the plane. The initial distance between P and the pulley is such that, in the subsequent motion, P does not reach the pulley.

d Show that the greatest height, in metres, reached by P above the plane is $1.26 g$. **(7 marks)**

E/P **3** Two particles A and B have masses m kg and 3 kg respectively, where $m > 3$. The particles are connected by a light inextensible string which passes over a smooth, fixed pulley. Initially A is 2.5 m above horizontal ground. The particles are released from rest with the string taut and the hanging parts of the string vertical, as shown in the figure. After A has been descending for 1.25 s, it strikes the ground. Particle A reaches the ground before B has reached the pulley.

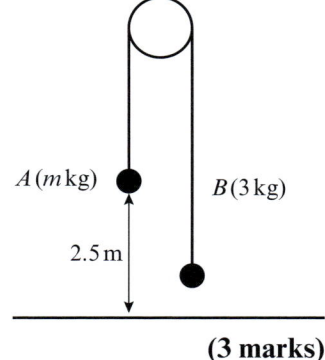

a Show that the acceleration of B as it ascends is 3.2 m s^{-2}.

(3 marks)

b Find the tension in the string as A descends. **(3 marks)**

c Show that $m = \frac{65}{11}$. **(4 marks)**

d State how you have used the information that the string is inextensible. **(1 mark)**

When A strikes the ground it does not rebound and the string becomes slack. Particle B then moves freely under gravity, without reaching the pulley, until the string becomes taut again.

e Find the time between the instant when A strikes the ground and the instant when the string becomes taut again. **(6 marks)**

4 Two particles A and B, of masses 5 kg and 3 kg respectively, are connected by a light inextensible string. Particle A lies on a rough horizontal table and the string passes over a small smooth pulley which is fixed at the edge of the table. Particle B hangs freely. The friction between A and the table is 24.5 N. The system is released from rest. Find:

a the acceleration of the system
b the tension in the string
c the magnitude of the force exerted on the pulley by the string.

E 5 A box P of mass 2.5 kg rests on a rough horizontal table and is attached to one end of a light inextensible string. The string passes over a small smooth pulley fixed at the edge of the table. The other end of the string is attached to a sphere Q of mass 1.5 kg which hangs freely below the pulley. The magnitude of the frictional force between P and the table is k N. The system is released from rest with the string taut. After release, Q descends a distance of 0.8 m in 0.75 s.

a Modelling P and Q as particles:
  i   calculate the acceleration of Q (3 marks)
  ii show that the tension in the string is 10.4 N (to 3 s.f.) (4 marks)
  iii find the value of k. (3 marks)
b State how in your calculations you have used the information that the string is inextensible. (1 mark)

Chapter review 4

1 A motorcycle of mass 200 kg is moving along a level road. The motorcycle's engine provides a forward thrust of 1000 N. The total resistance is modelled as a constant force of magnitude 600 N.
  a Modelling the motorcycle as a particle, draw a force diagram to show the forces acting on the motorcycle.
  b Calculate the acceleration of the motorcycle.

2 A man of mass 86 kg is standing in a lift which is moving upward with constant acceleration $2 \, \text{m s}^{-2}$. Find the magnitude and direction of the force that the man is exerting on the floor of the lift.

P 3 A car of mass 800 kg is travelling along a straight horizontal road. A constant retarding force of F N reduces the speed of the car from $18 \, \text{m s}^{-1}$ to $12 \, \text{m s}^{-1}$ in 2.4 s. Calculate:
  a the value of F b the distance moved by the car in these 2.4 s.

E 4 A block of mass 0.8 kg is pushed along a rough horizontal floor by a constant horizontal force of magnitude 7 N. The speed of the block increases from $2 \, \text{m s}^{-1}$ to $4 \, \text{m s}^{-1}$ in a distance of 4.8 m. Calculate:
  a the magnitude of the acceleration of the block (3 marks)
  b the magnitude of the frictional force between the block and the floor. (3 marks)

5 A car of mass 1200 kg is moving along a level road. The car's engine provides a constant driving force. The motion of the car is opposed by a constant resistance.

Given that the car is accelerating at 2 m s^{-2}, and that the magnitude of the driving force is three times the magnitude of the resistance force, show that the magnitude of the driving force is 3600 N.

6 Forces of (3**i** + 2**j**) N and (4**i** − **j**) N act on a particle of mass 0.25 kg. Find the acceleration of the particle.

7 Forces of $\begin{pmatrix} 2 \\ -1 \end{pmatrix}$ N, $\begin{pmatrix} 3 \\ -1 \end{pmatrix}$ N and $\begin{pmatrix} a \\ -2b \end{pmatrix}$ N act on a particle of mass 2 kg causing it to accelerate at $\begin{pmatrix} 3 \\ 2 \end{pmatrix}$ m s^{-2}. Find the values of a and b.

8 A sled of mass 2 kg is initially at rest on a horizontal plane. It is acted upon by a force of (2**i** + 4**j**) N for 3 seconds. Giving your answers in surd form,
   a find the magnitude of acceleration
   b find the distance travelled in the 3 seconds.

9 In this question, **i** and **j** represent the unit vectors due east and due north respectively.
The forces (3a**i** + 4b**j**) N, (5b**i** + 2a**j**) N and (−15**i** − 18**j**) N act on a particle of mass 2 kg which is in equilibrium.
   a Find the values of a and b.
   b The force (−15**i** − 18**j**) N is removed. Work out:
      i the magnitude and direction of the resulting acceleration of the particle
      ii the distance travelled by the particle in the first 3 seconds of its motion.

(E/P) 10 A car is towing a trailer along a straight horizontal road by means of a horizontal tow-rope. The mass of the car is 1400 kg. The mass of the trailer is 700 kg. The car and the trailer are modelled as particles and the tow-rope as a light inextensible string. The resistances to motion of the car and the trailer are assumed to be constant and of magnitude 630 N and 280 N respectively. The driving force on the car, due to its engine, is 2380 N. Find:
   a the acceleration of the car (3 marks)
   b the tension in the tow-rope. (3 marks)

When the car and trailer are moving at 12 m s^{-1}, the tow-rope breaks. Assuming that the driving force on the car and the resistances to motion are unchanged:
   c find the distance moved by the car in the first 4 s after the tow-rope breaks. (6 marks)
   d State how you have used the modelling assumption that the tow-rope is inextensible. (1 mark)

(E/P) 11 A train of mass 2500 kg pushes a carriage of mass 1100 kg along a straight horizontal track. The engine is connected to the carriage by a shunt which is parallel to the direction of motion of the coupling. The horizontal resistances to motion of the train and the carriage have magnitudes R N and 500 N respectively. The engine of the train produces a constant horizontal driving force of magnitude 8000 N that causes the train and carriage to accelerate at 1.75 m s^{-2}.

a Show that the resistance to motion R acting on the train is 1200 N. **(2 marks)**

b Find the magnitude of the compression in the shunt. **(3 marks)**

The train must stop at the next station, so the driver reduces the force produced by the engine to zero and applies the brakes. The brakes produce a force on the train of magnitude 2000 N causing the engine and carriage to decelerate.

c Given that the resistances to motion are unchanged, find the magnitude of the thrust in the shunt. Give your answer correct to 3 s.f. **(7 marks)**

(P) **12** Particles P and Q, of masses $2m$ kg and m kg respectively, are attached to the ends of a light inextensible string which passes over a smooth fixed pulley. They both hang at a distance of 2 m above horizontal ground. The system is released from rest.

a Find the magnitude of the acceleration of the system.

b Find the speed of P as it hits the ground.

Given that particle Q does not reach the pulley:

c find the greatest height that Q reaches above the ground.

d State how you have used in your calculation:
  i the fact that the string is inextensible   **ii** the fact that the pulley is smooth.

(E/P) **13** Two particles have masses 3 kg and m kg, where $m < 3$. They are attached to the ends of a light inextensible string. The string passes over a smooth fixed pulley. The particles are held in position with the string taut and the hanging parts of the string vertical, as shown. The particles are then released from rest. The initial acceleration of each particle has magnitude $\frac{3}{7}g$. Find:

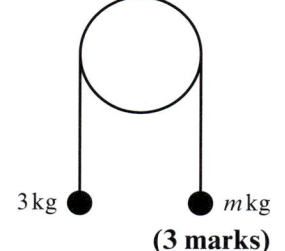

a the tension in the string immediately after the particles are released **(3 marks)**

b the value of m. **(3 marks)**

(E/P) **14** A block of wood A of mass 0.5 kg rests on a rough horizontal table and is attached to one end of a light inextensible string. The string passes over a small smooth pulley P fixed at the edge of the table. The other end of the string is attached to a ball B of mass 0.8 kg which hangs freely below the pulley, as shown in the figure. The resistance to motion of A from the rough table has a constant magnitude F N.

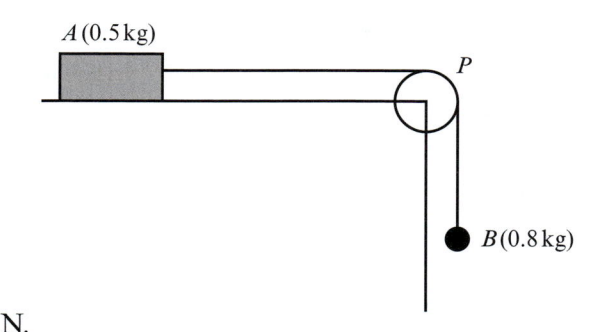

The system is released from rest with the string taut. After release, B descends a distance of 0.4 m in 0.5 s. Modelling A and B as particles, calculate:

a the acceleration of B **(3 marks)**

b the tension in the string **(4 marks)**

c the value of F. **(3 marks)**

d State how in your calculations you have used the information that the string is inextensible. **(1 mark)**

E/P 15 Two particles P and Q have masses 0.5 kg and 0.4 kg respectively. The particles are attached to the ends of a light inextensible string. The string passes over a small smooth pulley which is fixed above a horizontal floor. Both particles are held, with the string taut, at a height of 2 m above the floor, as shown. The particles are released from rest and in the subsequent motion Q does not reach the pulley.

    a **i** Write down an equation of motion for P.   **(2 marks)**
        ii Write down an equation of motion for Q.   **(2 marks)**

    b Find the tension in the string immediately after the particles are released.   **(2 marks)**

    c Find the acceleration of P immediately after the particles are released.   **(2 marks)**

When the particles have been moving for 0.2 s, the string breaks.

    d Find the further time that elapses until Q hits the floor.   **(9 marks)**

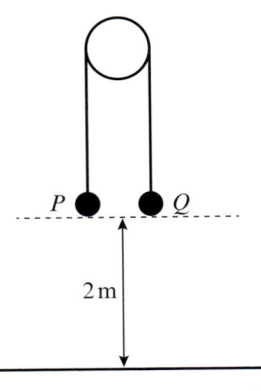

Challenge

In this question, **i** and **j** are the unit vectors due east and due north respectively.

Two boats start from rest at different points on the south bank of a river. The current in the river provides a constant force of magnitude 3**i** N on both boats.

The motor on boat A provides a thrust of $(-7\mathbf{i} + 2\mathbf{j})$ N and the motor on boat B provides a thrust of $(k\mathbf{i} + \mathbf{j})$ N. Given that the boats are accelerating in perpendicular directions, find the value of k.

Summary of key points

1. **Newton's first law of motion** states that an object at rest will stay at rest, and that an object moving with constant velocity will continue to move with constant velocity unless an unbalanced force acts on the object.

2. A **resultant** force acting on an object will cause the object to **accelerate in the same direction** as the resultant force.

3. You can find the resultant of two or more forces given as vectors by adding the vectors.

4. **Newton's second law of motion** states that the force needed to accelerate a particle is equal to the product of the mass of the particle and the acceleration produced: $\mathbf{F} = m\mathbf{a}$.

5. You can use $\mathbf{F} = m\mathbf{a}$ to solve problems involving vector forces acting on particles.

6. You can solve problems involving connected particles by considering the particles separately or, if they are moving in the same straight line, as a single particle.

7. **Newton's third law of motion** states that for every action there is an equal and opposite reaction.

Review exercise

1 The figure shows the velocity–time graph of a cyclist moving on a straight road over a 7 s period. The sections of the graph from $t = 0$ to $t = 3$, and from $t = 3$ to $t = 7$, are straight lines. The section from $t = 3$ to $t = 7$ is parallel to the t-axis.

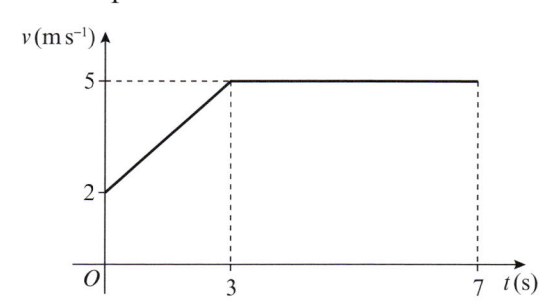

State what can be deduced about the motion of the cyclist from the fact that:

a the graph from $t = 0$ to $t = 3$ is a straight line (1)

b the graph from $t = 3$ to $t = 7$ is parallel to the t-axis. (1)

c Find the distance travelled by the cyclist during this 7 s period. (4)

← Mechanics 1 Section 2.2

2 A train stops at two stations 7.5 km apart. Between the stations it takes 75 s to accelerate uniformly to a speed of $24 \, \text{m s}^{-1}$, then travels at this speed for a time T seconds before decelerating uniformly for the final 0.6 km.

a Draw a velocity–time graph to illustrate this journey. (3)

Hence, or otherwise, find:

b the deceleration of the train during the final 0.6 km (3)

c the value of T (5)

d the total time for the journey. (4)

← Mechanics 1 Sections 2.2, 2.3

3 An electric train starts from rest at a station A and moves along a straight level track. The train accelerates uniformly at $0.4 \, \text{m s}^{-2}$ to a speed of $16 \, \text{m s}^{-1}$. This speed is then maintained for a distance of 2000 m. Finally, the train retards uniformly for 20 s before coming to rest at a station B. For this journey from A to B:

a find the total time taken (5)

b find the distance from A to B (5)

c sketch the displacement–time graph, showing clearly the shape of the graph for each stage of the journey. (3)

← Mechanics 1 Sections 2.1, 2.3

4 A small ball is projected vertically upward from a point A. The greatest height reached by the ball is 40 m above A. Calculate:

a the speed of projection (3)

b the time between the instant that the ball is projected and the instant it returns to A. (3)

← Mechanics 1 Sections 2.4, 2.5, 2.6

5 A ball is projected vertically upward and takes 3 seconds to reach its highest point. At time t seconds, the ball is 39.2 m above its point of projection. Find the possible values of t. (5)

← Mechanics 1 Sections 2.4, 2.5, 2.6

6 A light object is acted upon by a horizontal force of p N and a vertical force of q N as shown in the diagram.

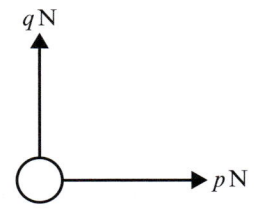

The resultant of the two forces has a magnitude of $\sqrt{40}$ N which acts in the direction of 30° to the horizontal. Calculate the value of p and the value of q.

← Mechanics 1 Sections 4.1, 4.2

(E) 7 A car of mass 750 kg, moving along a level straight road, has its speed reduced from 25 m s^{-1} to 15 m s^{-1} by brakes which produce a constant retarding force of 2250 N. Calculate the distance travelled by the car as its speed is reduced from 25 m s^{-2} to 15 m s^{-1}. **(5)**

← Mechanics 1 Sections 2.4, 4.3

(E) 8 An engine of mass 25 tonnes pulls a truck of mass 10 tonnes along a railway line. The resistances to the motion of the engine and the truck are modelled as constant and of magnitude 50 N per tonne. When the train is travelling horizontally, the tractive force (i.e. the force used to pull something) exerted by the engine is 26 kN. Modelling the engine and the truck as particles and the coupling that joins the engine and the truck as a light horizontal rod, calculate:

  a the acceleration of the engine and the truck **(4)**
  b the tension in the coupling. **(3)**
  c State how in your calculations you have used the information that:
      i the engine and the truck are particles
      ii the coupling is a light horizontal rod. **(2)**

← Mechanics 1 Sections 1.1, 1.2, 4.3, 4.5

(E/P) 9 A ball is projected vertically upward with a speed u m s^{-1} from a point A, which is 1.5 m above the ground. The ball moves freely under gravity until it reaches the ground. The greatest height attained by the ball is 25.6 m above A.

  a Show that $u = 22.4$. **(3)**

The ball reaches the ground T seconds after it has been projected from A.

  b Find, to three significant figures, the value of T. **(3)**

The ground is soft and the ball sinks 2.5 cm into the ground before coming to rest. The mass of the ball is 0.6 kg. The ground is assumed to exert a constant resistive force of magnitude F newtons.

  c Find, to three significant figures, the value of F. **(4)**
  d Sketch a velocity–time graph for the entire motion of the ball, showing the values of t at any points where the graph intercepts the horizontal axis. **(4)**
  e State one physical factor which could be taken into account to make the model used in this question more realistic. **(1)**

← Mechanics 1 Sections 1.1, 1.2, 2.5, 4.3, 4.4

(E) 10 A particle A, of mass 0.8 kg, resting on a smooth horizontal table, is connected to a particle B, of mass 0.6 kg, which is 1 m from the ground, by a light inextensible string passing over a small smooth pulley at the edge of the table. The particle A is more than 1 m from the edge of the table. The system is released from rest with the horizontal part of the string perpendicular to the edge of the table, the hanging parts vertical and the string taut.

Calculate:

  a the acceleration of A **(5)**
  b the tension in the string **(1)**
  c the speed of B when it hits the ground **(3)**
  d the time taken for B to reach the ground. **(3)**
  e The string in this question is described as being 'light'.
      i Write down what you understand by this description.
      ii State how you have used the fact that the string is light in your answer to parts **a** and **b**. **(2)**

← Mechanics 1 Sections 1.1, 1.2, 2.5, 4.6

REVIEW EXERCISE 1

11 Two particles P and Q have mass 0.6 kg and 0.2 kg respectively. The particles are attached to the ends of a light inextensible string. The string passes over a small smooth pulley which is fixed above a horizontal floor. Both particles are held, with the string taut, at a height of 1 m above the floor. The particles are released from rest and in the subsequent motion Q does not reach the pulley.

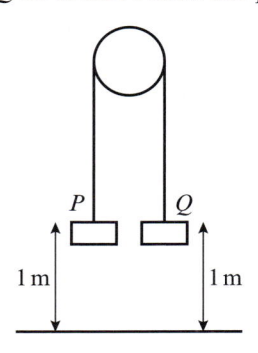

a Find the tension in the string immediately after the particles are released. **(6)**

b Find the acceleration of P immediately after the particles are released. **(2)**

When the particles have been moving for 0.4 s, the string breaks.

c Find the further time that elapses until P hits the floor. **(9)**

d State how in your calculations you have used the information that the string is inextensible. **(1)**

← Mechanics 1 Sections 2.5, 4.4, 4.6

12 A trailer of mass 600 kg is attached to a car of mass 900 kg by means of a light inextensible tow-bar. The car tows the trailer along a horizontal road. The resistances to motion of the car and trailer are 300 N and 150 N respectively.

a Given that the acceleration of the car and trailer is 0.4 m s^{-2}, calculate:

i the tractive force exerted by the engine of the car

ii the tension in the tow-bar. **(6)**

b Given that the magnitude of the force in the tow-bar must not exceed 1650 N, calcluate the greatest possible deceleration of the car. **(3)**

← Mechanics 1 Sections 4.1, 4.3, 4.5

13 A boy sits on a box in a lift. The mass of the boy is 45 kg, the mass of the box is 20 kg and the mass of the lift is 1050 kg. The lift is being raised vertically by a vertical cable which is attached to the top of the lift. The lift is moving upward and has constant deceleration of 2 m s^{-2}. By modelling the cable as being light and inextensible, find:

a the tension in the cable **(3)**

b the magnitude of the force exerted on the box by the boy **(3)**

c the magnitude of the force exerted on the box by the lift. **(3)**

← Mechanics 1 Sections 1.1, 1.2, 4.1, 4.4, 4.5

14 Two forces $\mathbf{F}_1 = (2\mathbf{i} + 3\mathbf{j})$ N and $\mathbf{F}_2 = (\lambda\mathbf{i} + \mu\mathbf{j})$ N, where λ and μ are scalars, act on a particle. The resultant of the two forces is $\mathbf{R}$, where $\mathbf{R}$ is parallel to the vector $\mathbf{i} + 2\mathbf{j}$.

a Find, to the nearest degree, the acute angle between the line of action of $\mathbf{R}$ and the vector $\mathbf{i}$. **(2)**

b Show that $2\lambda - \mu + 1 = 0$. **(5)**

c Given that the direction of $\mathbf{F}_2$ is parallel to $\mathbf{j}$, find the magnitude of $\mathbf{R}$, to 3 s.f. **(4)**

← Mechanics 1 Sections 3.5, 4.2

15 A force $\mathbf{R}$ acts on a particle, where $\mathbf{R} = (7\mathbf{i} + 16\mathbf{j})$ N.

Calculate:

a the magnitude of $\mathbf{R}$, giving your answer to one decimal place **(2)**

b the angle between the line of action of $\mathbf{R}$ and $\mathbf{i}$, giving your answer to the nearest degree. **(2)**

The force **R** is the resultant of two forces **P** and **Q**. The line of action of **P** is parallel to the vector $(\mathbf{i} + 4\mathbf{j})$ and the line of action of **Q** is parallel to the vector $(\mathbf{i} + \mathbf{j})$.

c Determine the forces **P** and **Q**, expressing each in terms of **i** and **j**. **(6)**

← Mechanics 1 Sections 3.5, 4.4

(E/P) **16** At noon, Rich has position vector $(\mathbf{i} - 6\mathbf{j})$ km and Dev has position vector $(9\mathbf{i} + 2\mathbf{j})$ km.

a Work out the distance between Rich and Dev. **(3)**

Rich moves with constant velocity $(\mathbf{i} + 6\mathbf{j})$ km h^{-1} and Dev moves with constant velocity $(-3\mathbf{i} + 2\mathbf{j})$ km h^{-1}.

b Show that Rich and Dev meet, and work out the time at which this occurs. **(8)**

c Find the position vector of the point at which they meet. **(2)**

← Mechanics 1 Section 3.6

(E) **17** Two forces, $\mathbf{F}_1$ and $\mathbf{F}_2$, act on a particle.
$\mathbf{F}_1 = (2\mathbf{i} - 5\mathbf{j})$ newtons
$\mathbf{F}_2 = (\mathbf{i} + \mathbf{j})$ newtons
The resultant force **R** acting on the particle is given by $\mathbf{R} = \mathbf{F}_1 + \mathbf{F}_2$.

a Calculate the magnitude of **R** in newtons. **(3)**

A third force, $\mathbf{F}_3$, begins to act on the particle, where $\mathbf{F}_3 = k\mathbf{j}$ newtons and k is a positive constant. The new resultant force is given by $\mathbf{R}_{new} = \mathbf{F}_1 + \mathbf{F}_2 + \mathbf{F}_3$.

b Given that the angle between the line of action of $\mathbf{R}_{new}$ and the vector **i** is 45 degrees, find the value of k. **(3)**

← Mechanics 1 Section 3.6

(E/P) **18** A helicopter takes off from its starting position O and travels 100 km on a bearing of 060°. It then travels 30 km due east before landing at point A. Given that the position vector of A relative to O is $(m\mathbf{i} + n\mathbf{j})$ km, find the exact values of m and n. **(4)**

← Mechanics 1 Sections 3.5, 4.2

(E/P) **19** At the very end of a race, Boat A has a position vector of $(-65\mathbf{i} + 180\mathbf{j})$ m and Boat B has a position vector of $(100\mathbf{i} + 120\mathbf{j})$ m. The finish line has a position vector of $10\mathbf{i}$ km.

a Show that Boat B is closer to the finish line than Boat A. **(2)**

Boat A is travelling at a constant velocity of $(2.5\mathbf{i} - 6\mathbf{j})$ m s^{-1} and Boat B is travelling at a constant velocity of $(-3\mathbf{i} - 4\mathbf{j})$ m s^{-1}.

b Calculate the speed of each boat. Hence, or otherwise, determine the result of the race. **(4)**

← Mechanics 1 Section 3.6

REVIEW EXERCISE 1

Challenge

1. A tram starts from rest at station A and travels in a straight line toward station B. It accelerates uniformly for t_1 seconds, covering a distance of 1750 m. It then travels at a constant speed v m s^{-1} for t_2 seconds, covering a distance of 17 500 m. The tram then decelerates for t_3 seconds and comes to rest at station B.

   Given that the total time for the journey is 7 minutes and $3t_1 = 4t_3$, find t_1, t_2 and t_3 and the distance between station A and station B.

   ← Mechanics 1 Sections 2.4, 2.5

2. **In this question, use $g = 10$ m s^{-2}.**

   One end of a light inextensible string is attached to a block A of mass 5 kg. Block A is held at rest on a rough horizontal table. The motion of the block is subject to a resistance of 2 N. The string lies parallel to the table and passes over a smooth light pulley which is fixed at the top of the table. The other end of the string is attached to a light scale-pan which carries two blocks B and C, with block B on top of block C as shown. The mass of block B is 5 kg and the mass of block C is 10 kg.

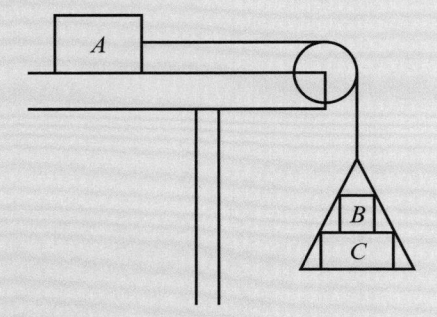

   The scale-pan hangs at rest and the system is released from rest. By modelling the blocks as particles, ignoring air resistance and assuming the motion is uninterrupted, find:

   a the acceleration of the scale-pan

   b the tension in the string

   c the magnitude of the force exerted on block B by block C

   d the magnitude of the force exerted on the pulley by the string.

   e State how you have used the information that the string is inextensible in your calculations.

   ← Mechanics 1 Sections 4.5, 4.6

5 FORCES AND FRICTION

4.4

Learning objectives

After completing this chapter you should be able to:
- Resolve forces into components → pages 85–90
- Use the triangle law to find a resultant force → pages 87–90
- Solve problems involving smooth or rough inclined planes → pages 90–93
- Understand friction and the coefficient of friction → pages 94–99
- Use $F \leq \mu R$ → pages 94–99

Prior knowledge check

1. A particle of mass 5 kg is acted on by two forces:
   $\mathbf{F}_1 = (8\mathbf{i} + 2\mathbf{j})$ N and $\mathbf{F}_2 = (-3\mathbf{i} + 8\mathbf{j})$ N.
   Find the acceleration of the particle in the form $(p\mathbf{i} + q\mathbf{j})$ m s^{-2}.
   ← Mechanics 1 Section 3.6

2. In the diagram below, calculate:
   a the length of the hypotenuse
   b the size of α.
   Give your answers correct to 2 d.p.

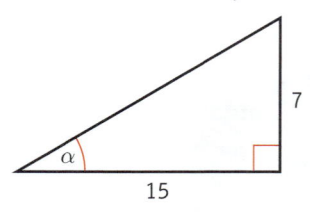

   ← International GCSE Mathematics

A car's braking distance is determined by its speed, the effectiveness of the brakes and the frictional force between the car's wheels and the road. In wet or icy conditions, friction is reduced so the braking distance is increased.

FORCES AND FRICTION CHAPTER 5

5.1 Resolving forces

- If a force is applied at an angle to the direction of motion you can resolve it to find the component of the force that acts in the direction of motion.

The book shown below is being dragged along the table by means of a force of magnitude F. The book is moving horizontally, and the angle between the force and the direction of motion is θ.

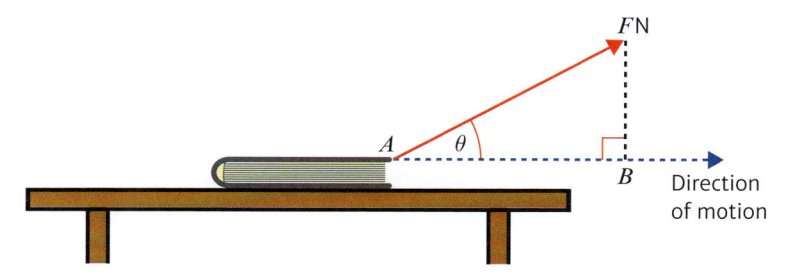

The effect of the force in the direction of motion is the length of AB. This is called the **component of the force in the direction of motion**. Using the rule for a right-angled triangle $\cos\theta = \dfrac{\text{adjacent}}{\text{hypotenuse}}$, you can see that the magnitude of the force in the direction of AB is $F \times \cos\theta$. Finding this value is called `**resolving**` the force in the direction of motion.

- The component of a force of magnitude F in a certain direction is $F\cos\theta$, where θ is the size of the angle between the force and the direction.

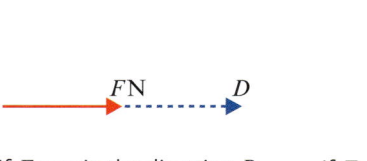

If F acts in the direction D, then the component of F in that direction is:
$F\cos 0° = F \times 1 = F$

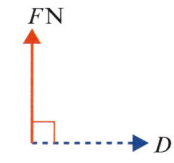

If F acts at a right angle to D, then the component of F in that direction is:
$F\cos 90° = F \times 0 = 0$

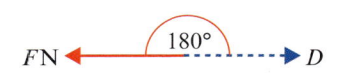

If F acts in the opposite direction to D, then the component of F in that direction is:
$F\cos 180° = F \times -1 = -F$

Example 1   SKILLS   PROBLEM-SOLVING

Find the component of each force in: **i** the x-direction   **ii** the y-direction.

iii Hence, write each force in the form $p\mathbf{i} + q\mathbf{j}$ where $\mathbf{i}$ and $\mathbf{j}$ are the unit vectors in the x and y directions respectively.

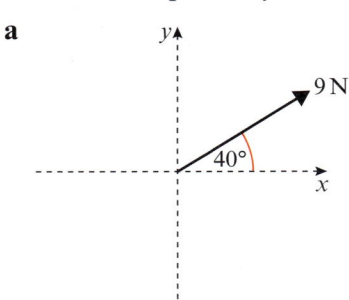

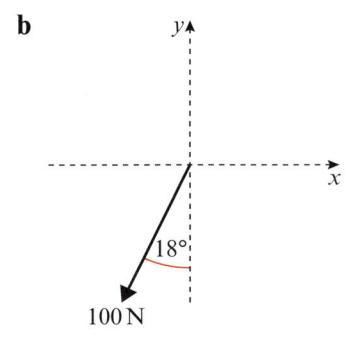

a 9 N at 40° above the x-axis

b 100 N at 18° below the negative y-axis (to the left)

a i $\theta = 40°$

Component in x-direction $= F\cos\theta$
$= 9 \times \cos 40°$
$= 6.89\,\text{N}$ (3 s.f.)

> Give your answers correct to three significant figures.

ii

$\theta = 90° - 40°$
$= 50°$

> Make sure you find the angle between the force and the direction you are resolving in.

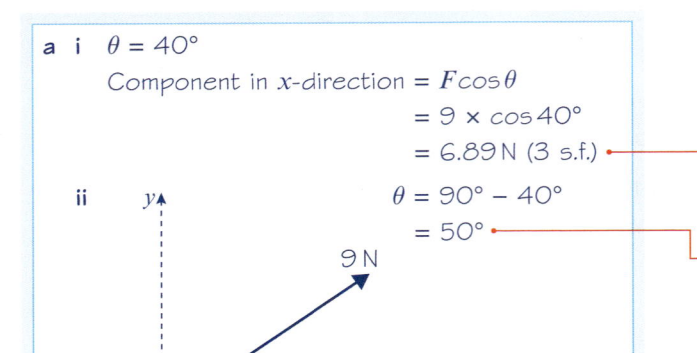

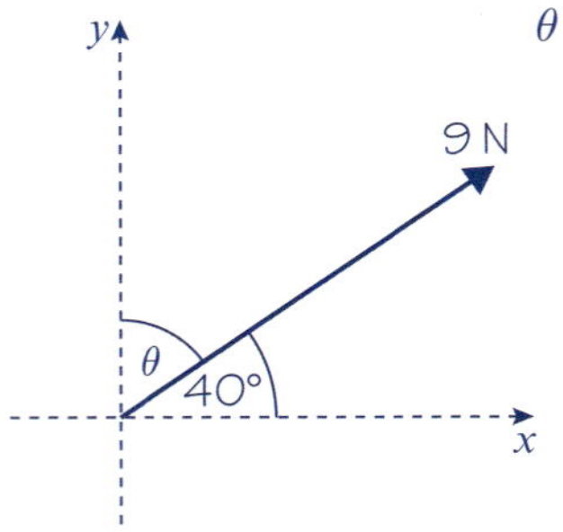

Component in y-direction $= F\cos\theta$
$= 9 \times \cos 50°$
$= 5.79\,\text{N}$ (3 s.f.)

> You could also use $F\sin 40°$ as $\sin 40° = \cos(90° - 40°) = \cos 50°$

iii $(6.89\mathbf{i} + 5.79\mathbf{j})\,\text{N}$

b i

$\theta = 90° + 18°$
$= 108°$

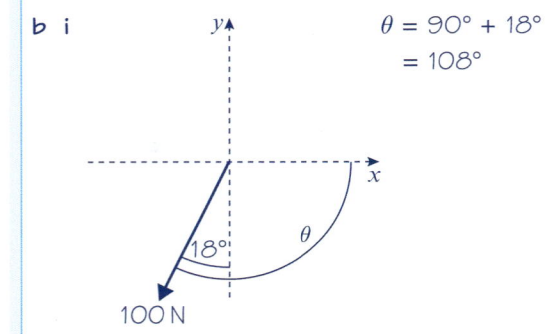

Component in x-direction $= F\cos\theta$
$= 100 \times \cos 108°$
$= -30.9\,\text{N}$ (3 s.f.)

> You get a negative answer because you are resolving in the positive x-direction. You could also resolve in the negative x-direction using $\theta = 90° - 18° = 72°$, then change the sign of your answer from positive to negative:

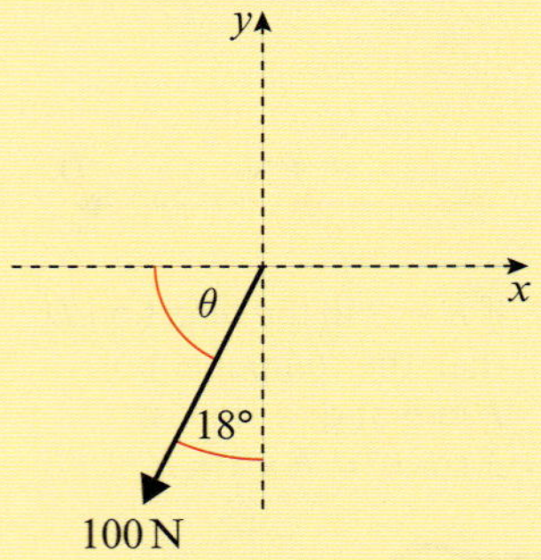

ii

$\theta = 180° - 18°$
$= 162°$

> You could use $\theta = 18°$ then change the sign of your answer from positive to negative: $-100\cos 18° = -95.1\,\text{N}$ (3 s.f.).

> You can measure θ in either the clockwise or the anticlockwise direction since $\cos\theta = \cos(360° - \theta)$.

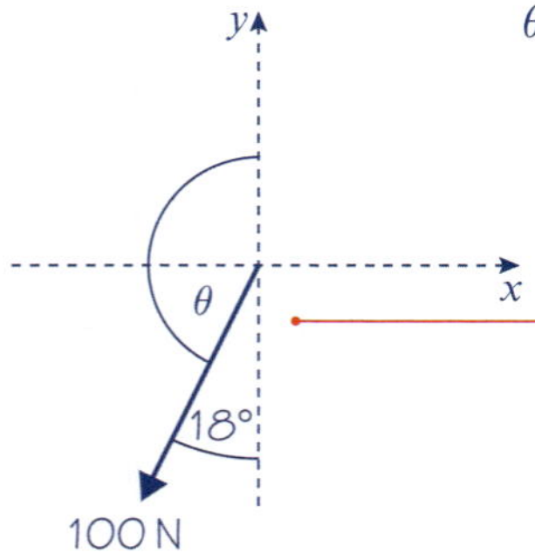

Component in y-direction $= F\cos\theta$
$= 100 \times \cos 162°$
$= -95.1\,\text{N}$ (3 s.f.)

iii $(-30.9\mathbf{i} - 95.1\mathbf{j})\,\text{N}$

FORCES AND FRICTION CHAPTER 5

Example 2 SKILLS PROBLEM-SOLVING

A box of mass 8 kg lies on a smooth horizontal floor. A force of 10 N is applied at an angle of 30° causing the box to accelerate horizontally along the floor.

a Work out the acceleration of the box.
b Calculate the normal reaction between the box and the floor.

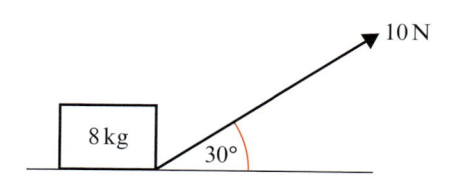

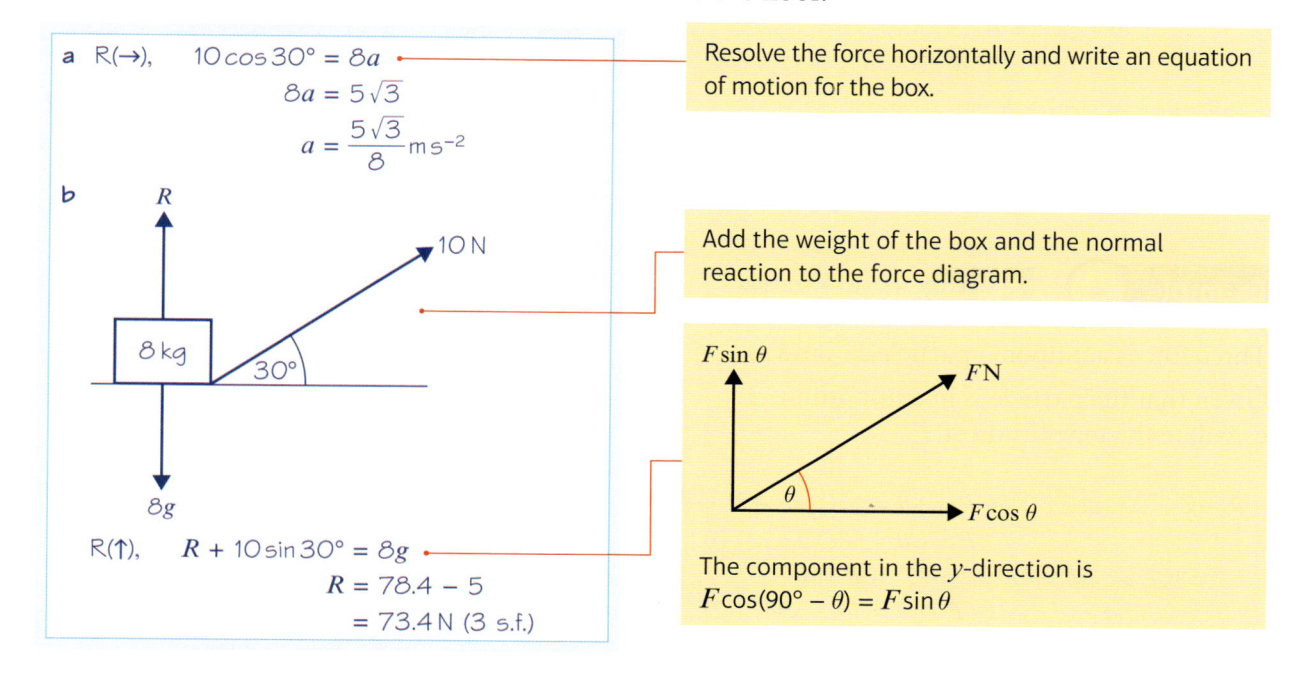

a $R(\rightarrow)$, $10\cos 30° = 8a$
$8a = 5\sqrt{3}$
$a = \dfrac{5\sqrt{3}}{8}\,\text{ms}^{-2}$

Resolve the force horizontally and write an equation of motion for the box.

b

Add the weight of the box and the normal reaction to the force diagram.

The component in the y-direction is $F\cos(90° - \theta) = F\sin\theta$

$R(\uparrow)$, $R + 10\sin 30° = 8g$
$R = 78.4 - 5$
$= 73.4\,\text{N}$ (3 s.f.)

You can use the **triangle law** of vector addition to find the resultant of two forces acting at an angle without resolving them into components.

Example 3 SKILLS CRITICAL THINKING

Two forces P and Q act on a particle as shown. P has a magnitude of 10 N and Q has a magnitude of 8 N. Work out the magnitude and direction of the resultant force.

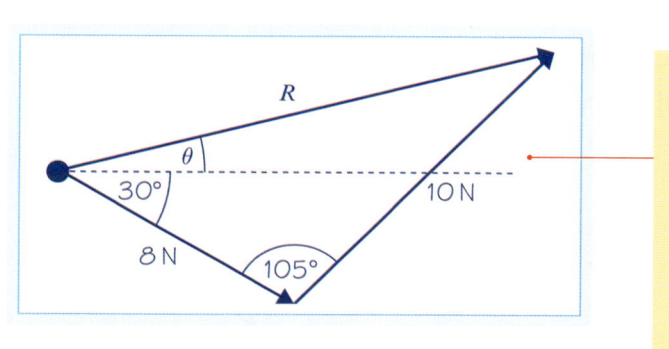

Use the triangle law for vector addition. The resultant force is the third side of a triangle formed by forces P and Q. You might need to use geometry to work out missing angles in the triangle.

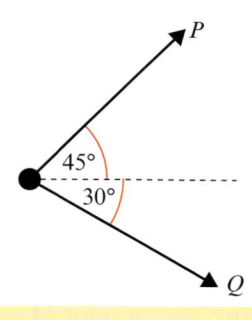

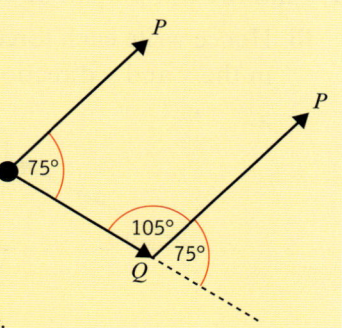

$R^2 = 8^2 + 10^2 - 2 \times 8 \times 10 \cos 105°$ — Use the cosine rule to calculate the magnitude of R.
$= 164 - 160 \cos 105° = 205.411...$
$R = 14.332... = 14.3 \text{ N} \text{ (3 s.f.)}$ — Use the sine rule to work out θ.

$\dfrac{\sin(\theta + 30°)}{10} = \dfrac{\sin 105°}{14.332...}$

Remember to use unrounded values in your calculations, then round your final answer.

$\sin(\theta + 30°) = \dfrac{10 \sin 105°}{14.332...} = 0.673...$

$\theta + 30° = 42.373...°$
$\theta = 12.4° \text{ (3 s.f.)}$

Use your diagram to check that your answer makes sense.

The resultant force R has a magnitude of 14.3 N and acts at an angle of 12.4° above the horizontal.

Online Explore the resultant of two forces using GeoGebra.

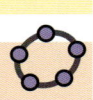

Example 4 SKILLS PROBLEM-SOLVING

Three forces act upon a particle as shown.
Given that the particle is in equilibrium, calculate the magnitude of P.

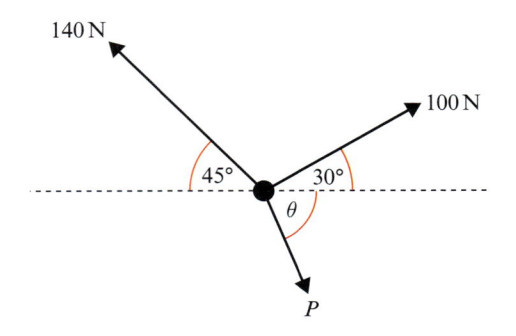

$R(\rightarrow)$, $100 \cos 30° + P \cos \theta = 140 \cos 45°$
$\quad P \cos \theta = 12.392...$ (1)
$R(\uparrow)$, $100 \sin 30° + 140 \sin 45° = P \sin \theta$
$\quad P \sin \theta = 148.994...$ (2)

$\dfrac{P \sin \theta}{P \cos \theta} = \dfrac{148.994...}{12.392...}$

$\tan \theta = 12.023...$
$\theta = 85.245...°$
$P \cos 85.2454...° = 12.392...$
$P = 150 \text{ N} \text{ (3 s.f.)}$

Resolve horizontally and vertically. You can solve these two equations simultaneously by dividing to eliminate P.

Problem-solving

You could also solve this problem by drawing a **triangle of forces**. The particle is in equilibrium, so the three forces will form a closed triangle.

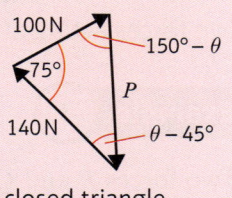

Exercise 5A SKILLS PROBLEM-SOLVING

1 Find the component of each force in: **i** the x-direction **ii** the y-direction.
  iii Hence write each force in the form $p\mathbf{i} + q\mathbf{j}$ where $\mathbf{i}$ and $\mathbf{j}$ are the unit vectors in the x and y directions respectively.

  a 12 N at 20° above positive x-axis
  b 5 N along negative y-axis
  c 8 N at 40° above negative x-axis
  d 6 N at 50° below negative y-axis (measured from negative y-axis, 50° toward negative x...)

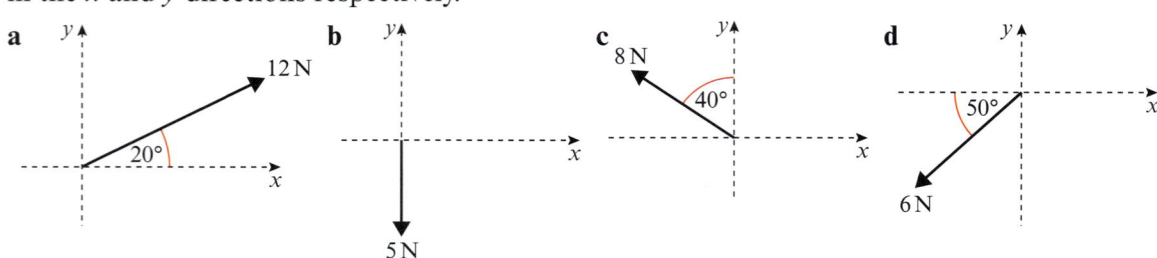

FORCES AND FRICTION CHAPTER 5

2 For each of the following systems of forces, find the sum of the components in:
  i the *x*-direction   **ii** the *y*-direction.

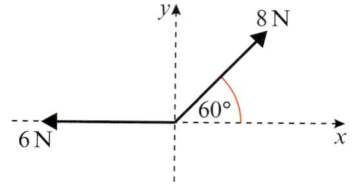

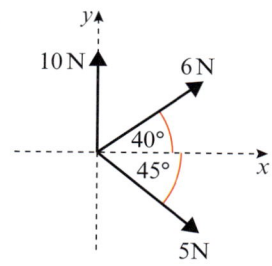

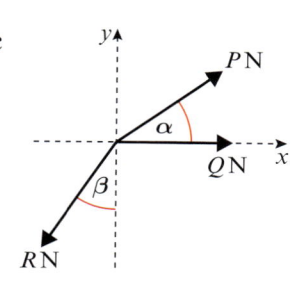

a (diagram: 6N along negative x-axis; 8N at 60° above positive x-axis)

b (diagram: 10N along positive y-axis; 6N at 40° above positive x-axis; 5N at 45° below positive x-axis)

c (diagram: P N at angle α above positive x-axis; Q N along positive x-axis; R N at angle β below positive x-axis)

3 Find the magnitude and direction of the resultant force acting on each of the particles shown below.

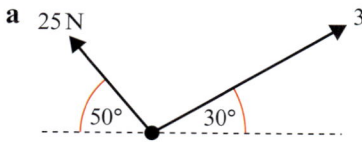

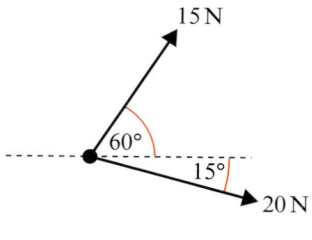

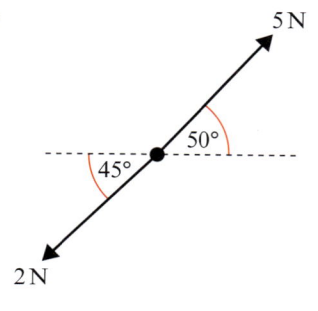

a 25 N at 50° above negative x-axis; 35 N at 30° above positive x-axis

b 15 N at 60° above positive x-axis; 20 N at 15° below positive x-axis

c 5 N at 50° above positive x-axis; 2 N at 45° below negative x-axis

P 4 Three forces act upon a particle as shown in the diagrams below.
Given that each particle is in equilibrium, calculate the magnitude of B and the value of θ.

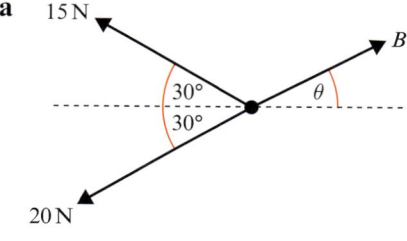

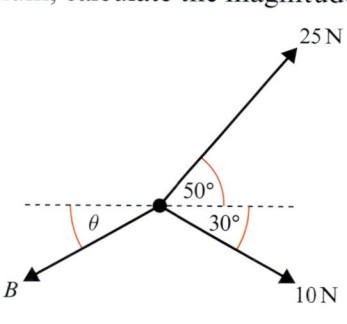

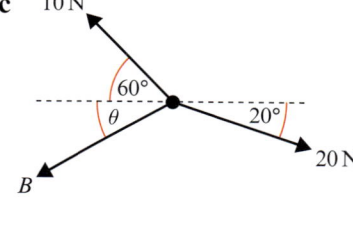

a 15 N at 30° above negative x-axis; 20 N at 30° below negative x-axis; B at angle θ above positive x-axis

b 25 N vertical upwards; B at 50° below negative x-axis (angle θ); 10 N at 30° below positive x-axis

c 10 N at 60° above negative x-axis; 20 N at 20° below positive x-axis; B at angle θ below negative x-axis

5 A box of mass 5 kg lies on a smooth horizontal floor. The box is pulled by a force of 2 N applied at an angle of 30° to the horizontal, causing the box to accelerate horizontally along the floor.

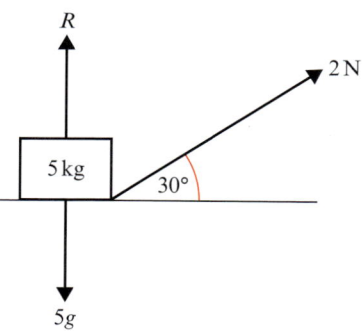

a Work out the acceleration of the box.

b Work out the normal reaction of the box with the floor.

(diagram: 5 kg box on ground with R upward, $5g$ downward, 2 N at 30° above horizontal)

E 6 A force P is applied to a box of mass 10 kg, causing the box to accelerate at $2\,\text{m s}^{-2}$ along a smooth, horizontal plane. Given that the force causing the acceleration is applied at 45° to the plane, work out the value of P. **(3 marks)**

E 7 A force of 20 N is applied to a box of mass m kg, causing the box to accelerate at $0.5\,\text{m s}^{-2}$ along a smooth, horizontal plane. Given that the force causing the acceleration is applied at 25° to the plane, work out the value of m. **(3 marks)**

E/P 8 A parachutist of mass 80 kg is attached to a parachute by two lines, each with tension T. The parachutist is falling with constant velocity, and experiences a resistance to motion due to air resistance equal to one quarter of her weight. Show that the tension in each line, T, is $20\sqrt{3}\,g$ N.

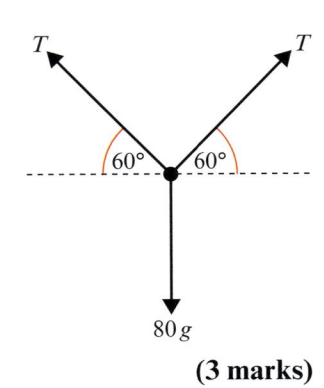

(3 marks)

E/P 9 A system of forces act upon a particle as shown in the diagram. The resultant force on the particle is $(2\sqrt{3}\mathbf{i} + 2\mathbf{j})$ N. Calculate the magnitudes of $\mathbf{F}_1$ and $\mathbf{F}_2$.

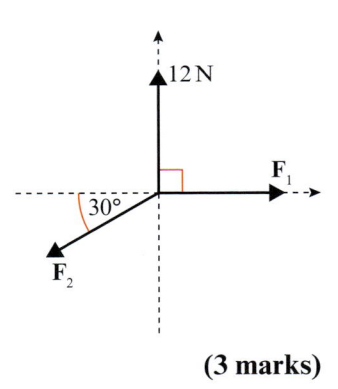

(3 marks)

> **Challenge**
> Two forces act upon a particle as shown in the diagram. The resultant force on the particle is $(3\mathbf{i} + 5\mathbf{j})$ N. Calculate the magnitudes of $\mathbf{F}_1$ and $\mathbf{F}_2$.

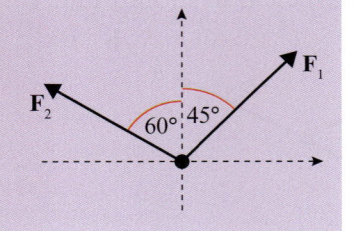

5.2 Inclined planes

Force diagrams may be used to model situations involving objects on **inclined** planes.

- **To solve problems involving inclined planes, it is usually easier to resolve forces parallel to and at right angles to the plane.**

Example 5   SKILLS   PROBLEM-SOLVING

A block of mass 10 kg slides down a smooth slope angled at 15° to the horizontal.
a Draw a force diagram to show all the forces acting on the block.
b Calculate the magnitude of the normal reaction of the slope on the block.
c Find the acceleration of the block.

FORCES AND FRICTION — CHAPTER 5

a

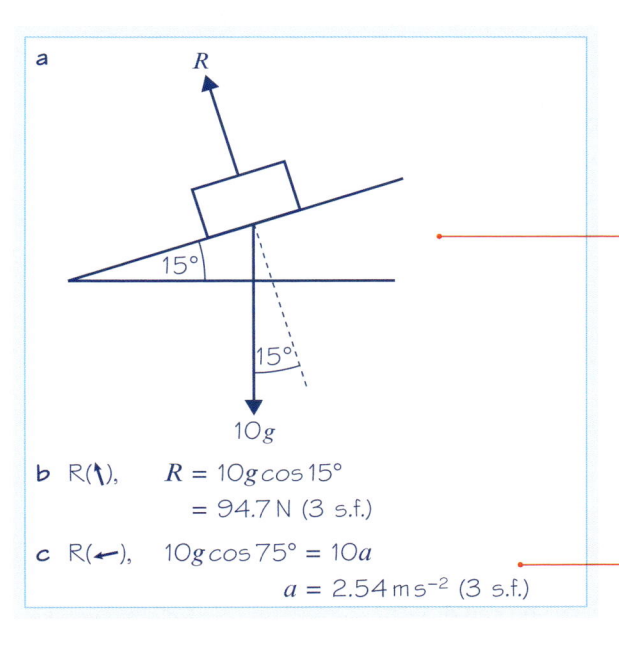

[Diagram: block on slope at 15°, with R perpendicular to slope and weight 10g vertically down; 15° angle between weight and perpendicular to slope]

b R($\nwarrow$), $R = 10g\cos 15°$
$= 94.7\,\text{N}$ (3 s.f.)

c R($\leftarrow$), $10g\cos 75° = 10a$
$a = 2.54\,\text{m s}^{-2}$ (3 s.f.)

Watch out The normal reaction force acts at right angles to the plane, not vertically.

Your working will be easier if you resolve at right angles to the plane. The weight of the block acts at an angle of 15° to this direction.

Notation The diagonal arrows, R($\nwarrow$) and R($\leftarrow$), show that you are resolving down the slope and perpendicular to the slope. You can also use R(//) to show resolution parallel to the slope and R($\perp$) to show resolution perpendicular to the slope.

Resolve down the slope and use $F = ma$.

Example 6 — SKILLS PROBLEM-SOLVING

A particle of mass m is pushed up a smooth slope by a force of magnitude $5g$ N acting at an angle of 60° to the slope, causing the particle to accelerate up the slope at $0.5\,\text{m s}^{-2}$. Find the mass of the particle.

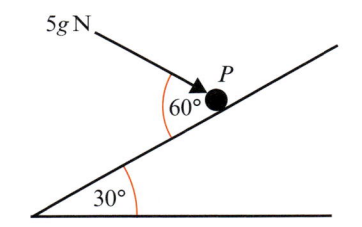

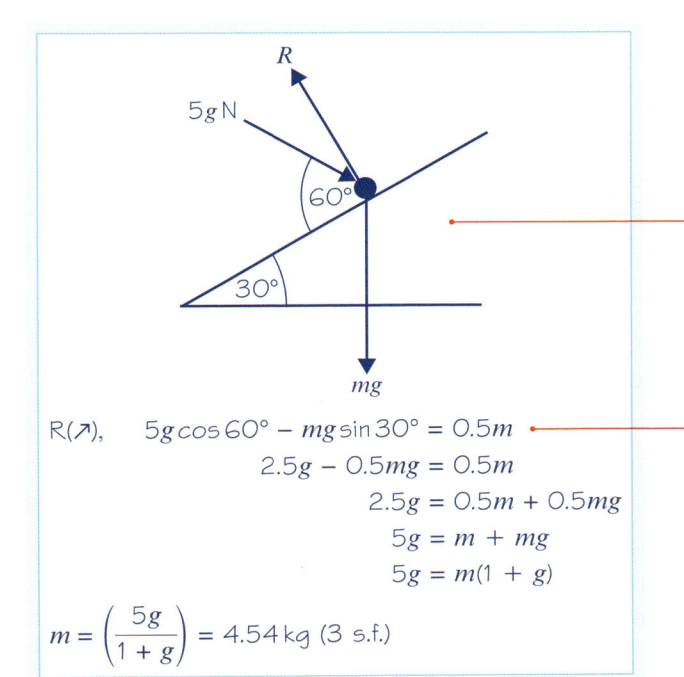

[Diagram: particle P on slope inclined at 30°, force 5g N applied at 60° above the slope]

[Diagram: forces on particle — R perpendicular to slope, 5g N at 60° to slope, mg vertically down, slope angle 30°]

R($\nearrow$), $5g\cos 60° - mg\sin 30° = 0.5m$
$2.5g - 0.5mg = 0.5m$
$2.5g = 0.5m + 0.5mg$
$5g = m + mg$
$5g = m(1 + g)$

$m = \left(\dfrac{5g}{1+g}\right) = 4.54\,\text{kg}$ (3 s.f.)

Draw a diagram to show all the forces acting on the particle.

Resolve up the slope, in the direction of the acceleration, and write an equation of motion for the particle.

Example 7 — SKILLS: PROBLEM-SOLVING

A particle P of mass 2 kg is moving on a smooth slope and is being acted on by a force of 4 N that acts parallel to the slope, as shown.

The slope is inclined at an angle α to the horizontal, where $\tan \alpha = \frac{3}{4}$. Work out the acceleration of the particle.

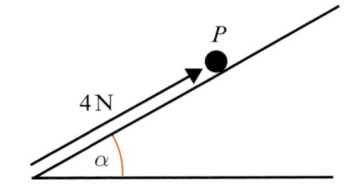

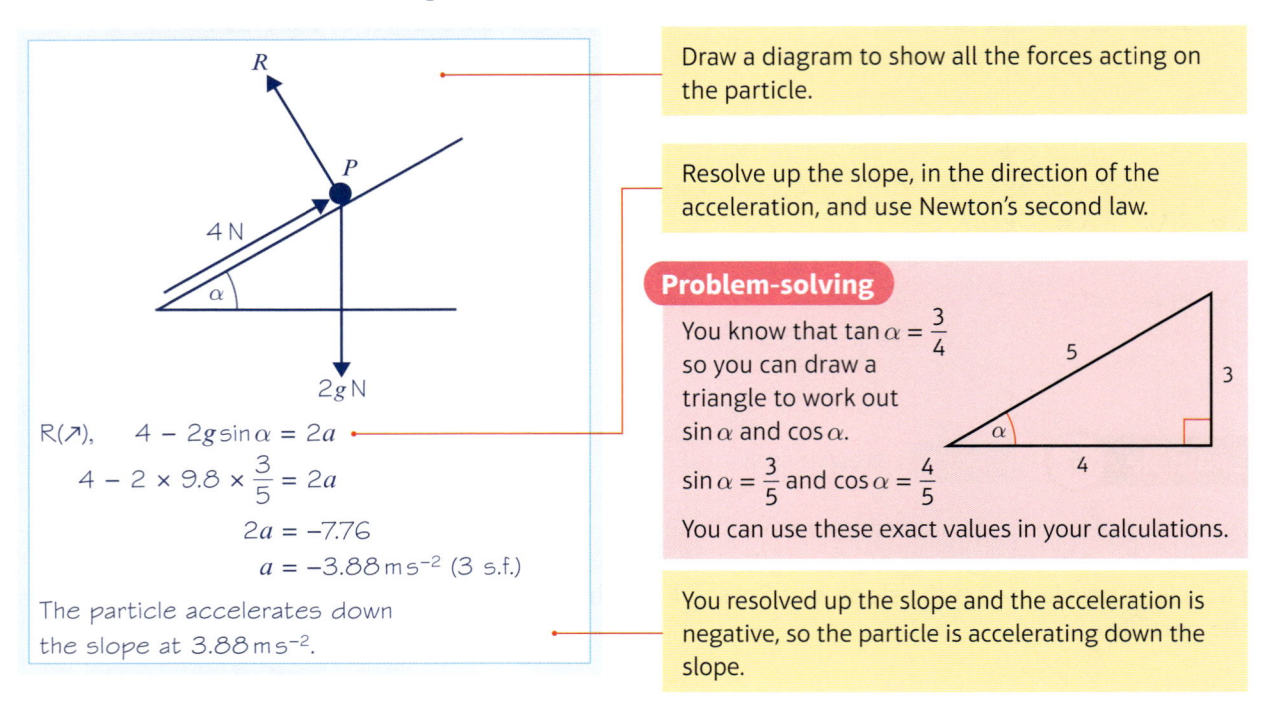

Draw a diagram to show all the forces acting on the particle.

Resolve up the slope, in the direction of the acceleration, and use Newton's second law.

Problem-solving

You know that $\tan \alpha = \frac{3}{4}$ so you can draw a triangle to work out $\sin \alpha$ and $\cos \alpha$.

$\sin \alpha = \frac{3}{5}$ and $\cos \alpha = \frac{4}{5}$

You can use these exact values in your calculations.

$R(\nearrow)$, $4 - 2g\sin\alpha = 2a$

$4 - 2 \times 9.8 \times \frac{3}{5} = 2a$

$2a = -7.76$

$a = -3.88 \, \text{m s}^{-2}$ (3 s.f.)

The particle accelerates down the slope at $3.88 \, \text{m s}^{-2}$.

You resolved up the slope and the acceleration is negative, so the particle is accelerating down the slope.

Exercise 5B — SKILLS: PROBLEM-SOLVING

1 A particle of mass 3 kg slides down a smooth plane that is inclined at 20° to the horizontal.
  a Draw a force diagram to represent all the forces acting on the particle.
  b Work out the normal reaction between the particle and the plane.
  c Find the acceleration of the particle.

2 A force of 50 N is pulling a particle of mass 5 kg up a smooth plane that is inclined at 30° to the horizontal. Given that the force acts parallel to the plane,
  a draw a force diagram to represent all the forces acting on the particle
  b work out the normal reaction between the particle and the plane
  c find the acceleration of the particle.

3 A particle of mass 0.5 kg is held at rest on a smooth slope that is inclined at an angle α to the horizontal. The particle is released. Given that $\tan \alpha = \frac{3}{4}$, calculate:
  a the normal reaction between the particle and the plane
  b the acceleration of the particle.

FORCES AND FRICTION CHAPTER 5 93

(E) **4** A force of 30 N is pulling a particle of mass 6 kg up a rough slope that is inclined at 15° to the horizontal. The force acts in the direction of motion of the particle and the particle experiences a constant resistance due to friction.

    a Draw a force diagram to represent all the forces acting on the particle. **(4 marks)**

Given that the particle is moving with constant speed,

    b calculate the magnitude of the resistance due to friction. **(5 marks)**

(E) **5** A particle of mass m kg is sliding down a smooth slope that is angled at 30° to the horizontal. The normal reaction between the plane and the particle is 5 N.

    a Calculate the mass m of the particle. **(3 marks)**

    b Calculate the acceleration of the particle. **(3 marks)**

(E/P) **6** A force of 30 N acts horizontally on a particle of mass 5 kg that rests on a smooth slope that is inclined at 30° to the horizontal as shown in the diagram. Find the acceleration of the particle.

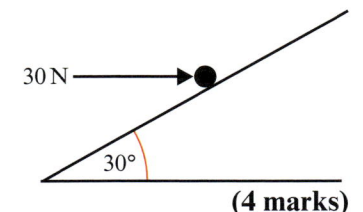

(4 marks)

(E/P) **7** A particle of mass 3 kg is moving on a rough slope that is inclined at 40° to the horizontal. A force of 6 N acts vertically upon the particle. Given that the particle is moving at a constant velocity, calculate the value of F, the constant resistance due to friction.

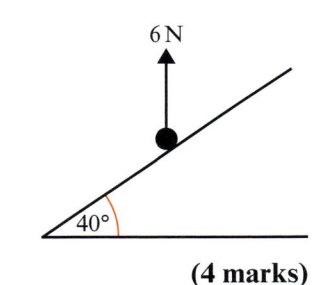

(4 marks)

(E/P) **8** A particle of mass m kg is pulled up a rough slope by a force of 26 N that acts at an angle of 45° to the slope. The particle experiences a constant frictional force of magnitude 12 N.

Given that $\tan \alpha = \dfrac{1}{\sqrt{3}}$ and that the acceleration of the particle is $1\,\text{m s}^{-2}$, show that $m = 1.08$ kg (3 s.f.).

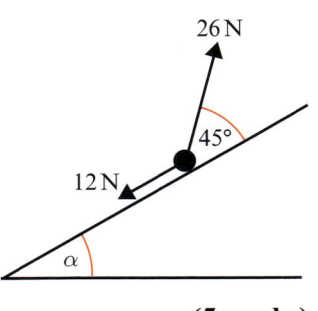

(5 marks)

Challenge

A particle is sliding down a smooth slope inclined at an angle θ to the horizontal, where $0 < \theta < 30°$. The angle of inclination of the slope is increased by 60°, and the magnitude of the acceleration of the particle increases from a to $4a$.

a Show that $\tan \theta = \dfrac{\sqrt{3}}{7}$.

b Hence find θ, giving your answer to 3 significant figures.

5.3 Friction

Friction is a force which opposes motion between two rough surfaces. It occurs when the two surfaces are moving relative to one another, or when there is a **tendency** (i.e. where something is likely to behave in a certain way) for them to move relative to one another.

This block is stationary. There is no horizontal force being applied, so there is no tendency for the block to move. There is no frictional force acting on the block.

This block is also stationary. There is a horizontal force being applied which is not sufficient to move the block. There is a tendency for the block to move, but it doesn't because the force of friction is equal and opposite to the force being applied.

As the applied force increases, the force of friction increases to prevent the block from moving. If the magnitude of the applied force exceeds a certain **maximum** or **limiting value**, the block will move. While the block moves, the force of friction will remain constant at its maximum value.

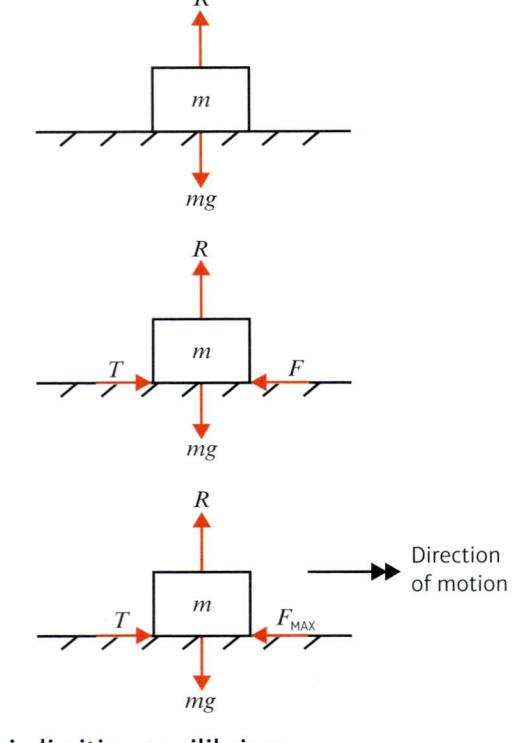

- When a particle is on the point of moving it is said to be in limiting equilibrium.

The limiting value of the friction depends on two things:
- the normal reaction R between the two surfaces in contact
- the roughness of the two surfaces in contact.

You can measure roughness using the **coefficient of friction**, which is represented by the letter μ (pronounced *myoo*). The rougher the two surfaces, the larger the value of μ. For smooth surfaces there is no friction and $\mu = 0$.

- The maximum or limiting value of the friction between two surfaces, F_{MAX}, is given by

   $F_{MAX} = \mu R$

   where μ is the coefficient of friction and R is the normal reaction between the two surfaces.

Example 8   SKILLS   PROBLEM-SOLVING

A particle of mass 5 kg is pulled along a rough horizontal surface by a horizontal force of magnitude 20 N. The coefficient of friction between the particle and the floor is 0.2. Calculate:
a the magnitude of frictional force
b the acceleration of the particle.

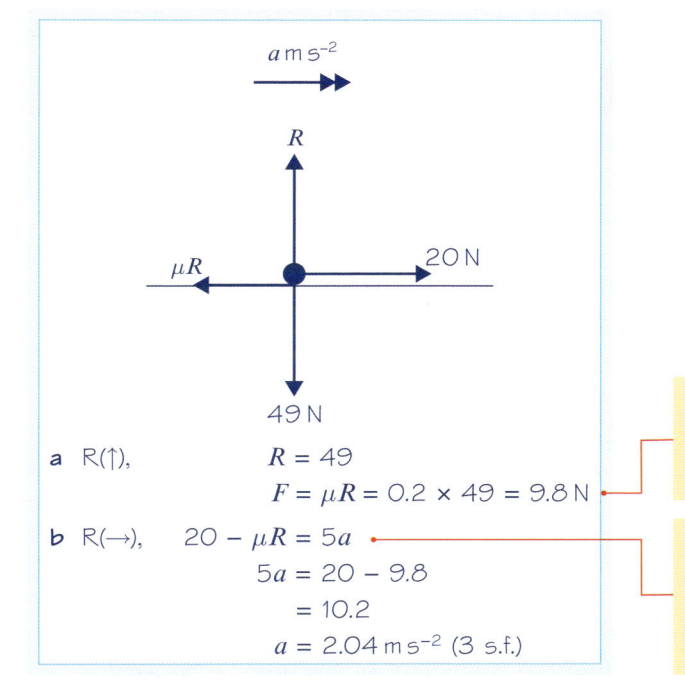

FORCES AND FRICTION CHAPTER 5

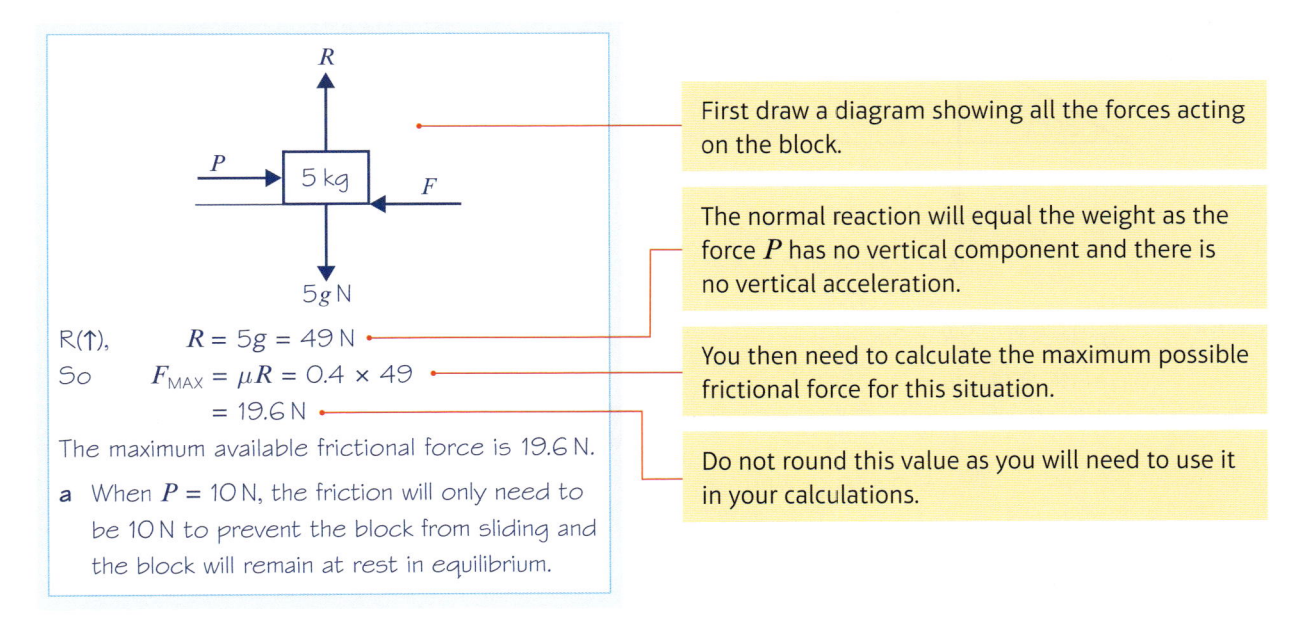

a R(↑), $R = 49$
          $F = \mu R = 0.2 \times 49 = 9.8\,\text{N}$

> For a moving particle, $F = F_{MAX}$ so you can use $F = \mu R$ to find the magnitude of the frictional force.

b R(→), $20 - \mu R = 5a$
          $5a = 20 - 9.8$
            $= 10.2$
          $a = 2.04\,\text{m s}^{-2}$ (3 s.f.)

> Write an equation of motion for the particle, resolving horizontally. Note that the frictional force always acts in a direction so as to **oppose** the motion of the particle.

Example 9   SKILLS PROBLEM-SOLVING

A block of mass 5 kg lies on rough horizontal ground. The coefficient of friction between the block and the ground is 0.4. A horizontal force P is applied to the block. Find the magnitude of the frictional force acting on the block and the acceleration of the block when the magnitude of P is:
a 10 N   **b** 19.6 N   **c** 30 N.

> First draw a diagram showing all the forces acting on the block.

> The normal reaction will equal the weight as the force P has no vertical component and there is no vertical acceleration.

R(↑), $R = 5g = 49\,\text{N}$
So $F_{MAX} = \mu R = 0.4 \times 49$
             $= 19.6\,\text{N}$

> You then need to calculate the maximum possible frictional force for this situation.

The maximum available frictional force is 19.6 N.

> Do not round this value as you will need to use it in your calculations.

a When $P = 10\,\text{N}$, the friction will only need to be 10 N to prevent the block from sliding and the block will remain at rest in equilibrium.

CHAPTER 5 — FORCES AND FRICTION

b When $P = 19.6\,\text{N}$, the friction will need to be at its maximum value of 19.6 N to prevent the block from sliding, and the block will remain at rest in **limiting equilibrium**.

c When $P = 30\,\text{N}$, the friction will be unable to prevent the block from sliding, and it will remain at its maximum value of 19.6 N. The block will accelerate from rest along the plane in the direction of P with acceleration a, where
$$30 - 19.6 = 5a$$
$$a = 2.08\,\text{m s}^{-2}\ (3\text{ s.f.})$$

> **Watch out** An object in limiting equilibrium can either be at rest or moving with constant velocity.

Example 10 — SKILLS: PROBLEM-SOLVING

A particle of mass 2 kg is sliding down a rough slope that is inclined at 30° to the horizontal. Given that the acceleration of the particle is $1\,\text{m s}^{-2}$, find the coefficient of friction, μ, between the particle and the slope.

> **Online** Explore this problem with different masses, slopes and frictional coefficients using GeoGebra.

Draw a diagram showing the weight, the frictional force and the normal reaction.

$R(\nwarrow)$, $\qquad R = 2g\cos 30°$
$\qquad\qquad\quad = 16.974\ldots$

$R(\nearrow)$, $\quad 2g\sin 30° - \mu R = 2a$

Use $F = ma$ to write an equation of motion for the particle.

$\qquad 9.8 - (16.974\ldots)\mu = 2$

$$\mu = \frac{7.8}{16.974\ldots}$$
$$= 0.460\ (3\text{ s.f.})$$

> **Watch out** Make sure you use the normal reaction, not the weight, when substituting into $F_{\text{MAX}} = \mu R$.

FORCES AND FRICTION — CHAPTER 5

Exercise 5C — SKILLS: PROBLEM-SOLVING

1 Each of the following diagrams shows a body of mass 5 kg lying initially at rest on rough horizontal ground. The coefficient of friction between the body and the ground is $\frac{1}{7}$. In each diagram, R is the normal reaction of the ground on the body and F is the frictional force exerted on the body. Any other forces applied to the body are as shown on the diagrams.

Hint The forces acting on the body can affect the magnitude of the normal reaction. In part **d** the normal reaction is $(5g + 14)$ N, so $F_{MAX} = \mu(5g + 14)$ N.

In each case:

i find the magnitude of F
ii state whether the body will remain at rest or accelerate from rest along the ground
iii find, when appropriate, the magnitude of this acceleration.

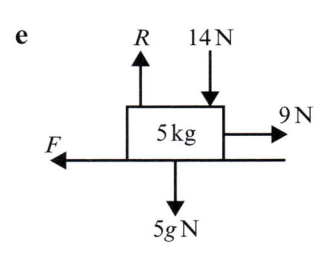

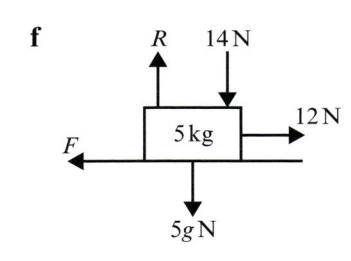

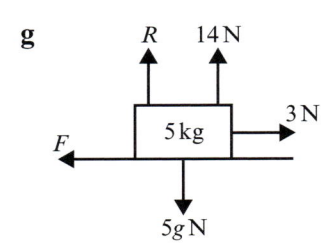

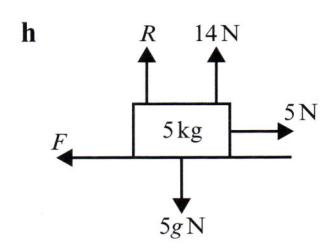

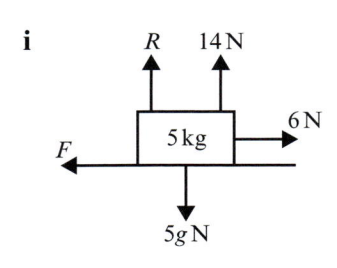

a 5 kg, applied force 3 N →
b 5 kg, applied force 7 N →
c 5 kg, applied force 12 N →
d 5 kg, 14 N downward, 6 N →
e 5 kg, 14 N downward, 9 N →
f 5 kg, 14 N downward, 12 N →
g 5 kg, 14 N upward, 3 N →
h 5 kg, 14 N upward, 5 N →
i 5 kg, 14 N upward, 6 N →
j 5 kg, 14 N at 30° above horizontal
k 5 kg, 28 N at 30° above horizontal
l 5 kg, 56 N at 45° above horizontal (pushing down), F to the left

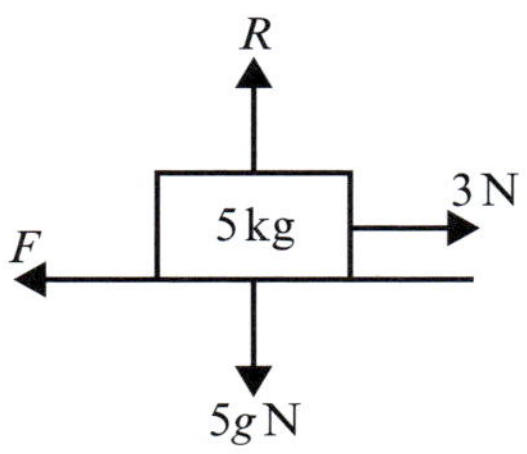

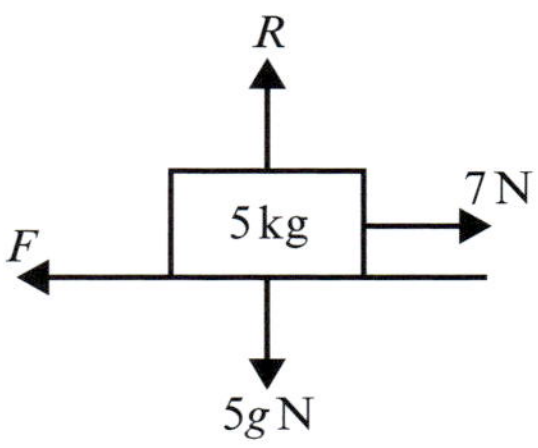

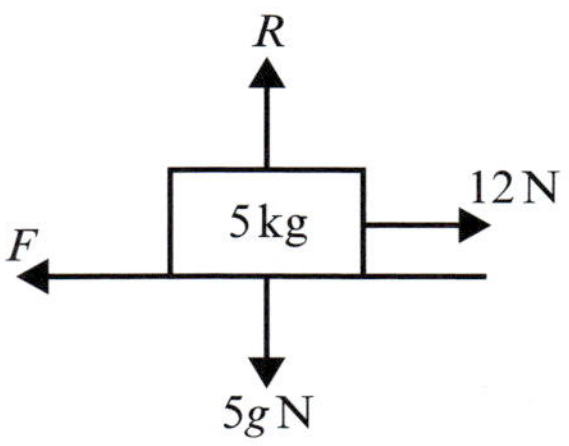

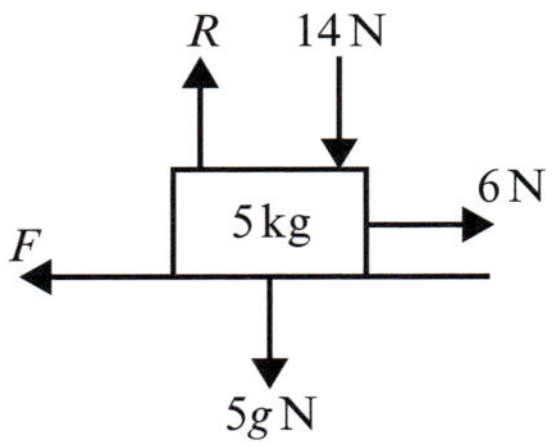

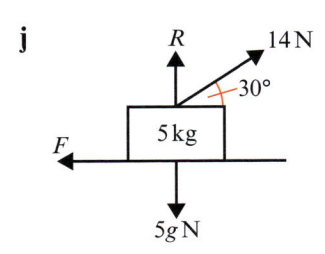

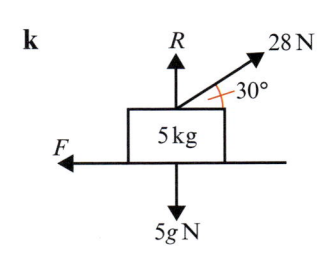

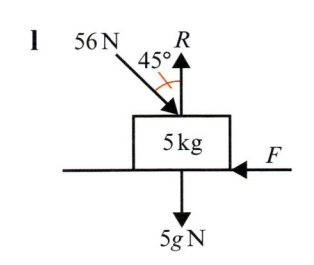

2 In each of the following diagrams, the forces cause the body of mass 10 kg to accelerate as shown along the rough horizontal plane. R is the normal reaction and F is the frictional force.

Find the normal reaction and the coefficient of friction in each case.

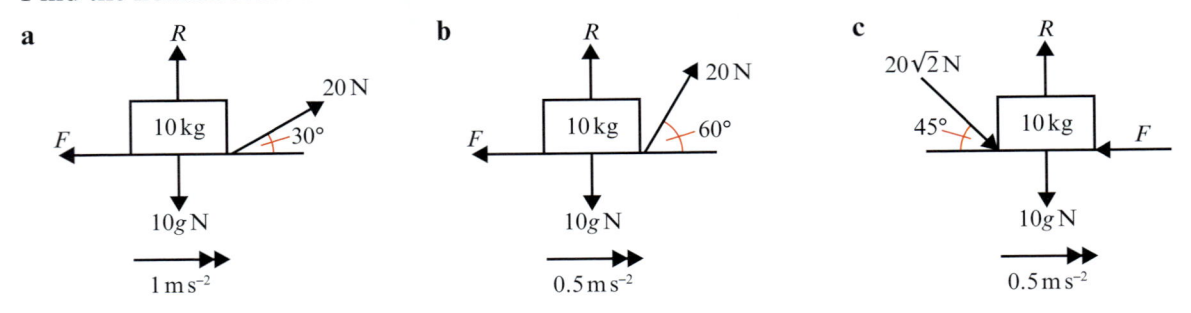

a Force 20 N at 30° above horizontal (pulling), 10 kg block, weight 10g N, acceleration 1 m s⁻².

b Force 20 N at 60° above horizontal (pulling), 10 kg block, weight 10g N, acceleration 0.5 m s⁻².

c Force 20√2 N at 45° above horizontal (pushing down from left), 10 kg block, weight 10g N, acceleration 0.5 m s⁻².

(E) 3 A particle of mass 0.5 kg is sliding down a rough slope that is angled at 15° to the horizontal. The acceleration of the particle is 0.25 m s⁻². Calculate the coefficient of friction between the particle and the slope. **(3 marks)**

(E) 4 A particle of mass 2 kg is sliding down a rough slope that is angled at 20° to the horizontal. A force of magnitude P acts parallel to the slope and is attempting to pull the particle up the slope. The acceleration of the particle is 0.2 m s⁻² down the slope and the coefficient of friction between the particle and the slope is 0.3. Find the value of P. **(4 marks)**

5 A particle of mass 5 kg is being pushed up a rough slope that is angled at 30° to the horizontal by a horizontal force P. Given that the coefficient of friction is 0.2 and the acceleration of the particle is 2 m s⁻², calculate the value of P.

(E/P) 6 A sledge of mass 10 kg is being pulled along a rough horizontal plane by a force P that acts at an angle of 45° to the horizontal. The coefficient of friction between the sledge and the plane is 0.1. Given that the sledge accelerates at 0.3 m s⁻², find the value of P. **(7 marks)**

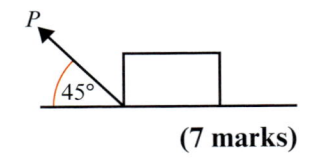

(E/P) 7 A train of mass m kg is travelling at 20 m s⁻¹ when it applies its brakes, causing the wheels to lock up. The train decelerates at a constant rate, coming to a complete stop in 30 seconds.

> **Problem-solving**
> Use the formulae for constant acceleration.
> ← Mechanics 1 Sections 2.4, 2.5

a By modelling the train as a particle, find the coefficient of friction between the railway track and the wheels of the train. **(6 marks)**

The train is no longer modelled as a particle, so that the effects of air resistance can be taken into account.

b State, with a reason, whether the value of the coefficient of friction calculated using this revised model would be greater than or less than the value calculated in **a**. **(2 marks)**

8 A box A of mass 2 kg is held at rest on a rough horizontal plane that is inclined at 30° to the horizontal. The coefficient of friction between the box and the plane is $\frac{1}{\sqrt{3}}$. Box A is connected to a second box B of mass 3 kg by a taut light inextensible string that passes over a smooth pulley. Calculate the acceleration of the system when box A is released from rest. **(10 marks)**

FORCES AND FRICTION CHAPTER 5

Challenge

A particle of mass m kg is sliding down a rough slope that is angled at α to the horizontal. The coefficient of friction between the particle and the slope is μ. Show that the acceleration of the particle is independent of its mass.

Chapter review 5

1 A box of mass 3 kg lies on a smooth horizontal floor. A force of 3 N is applied at an angle of 60° to the horizontal, causing the box to accelerate horizontally along the floor.
   a Find the magnitude of the normal reaction of the floor on the box.
   b Find the acceleration of the box.

2 A system of forces acts upon a particle as shown in the diagram. The resultant force on the particle is $(3\mathbf{i} + 2\mathbf{j})$ N. Calculate the magnitudes $\mathbf{F}_1$ and $\mathbf{F}_2$.

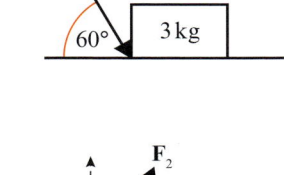

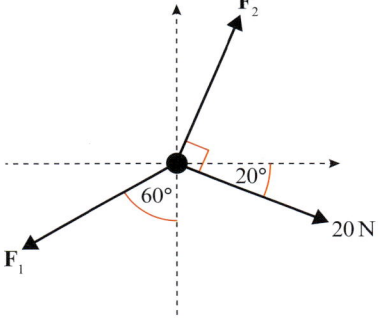

3 A force of 20 N is pulling a particle of mass 2 kg up a rough slope that is inclined at 45° to the horizontal. The force acts parallel to the slope, and the resistance due to friction is constant and has magnitude 4 N.
   a Draw a force diagram to represent all the forces acting on the particle.
   b Work out the normal reaction between the particle and the plane.
   c Show that the acceleration of the particle is 1.1 m s^{-2} (2 s.f.).

(E) 4 A particle of mass 5 kg sits on a smooth slope that is inclined at 10° to the horizontal. A force of 20 N acts on the particle at an angle of 20° to the plane, as shown in the diagram. Find the acceleration of the particle.
(5 marks)

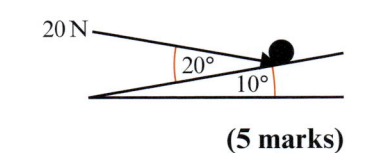

(E/P) 5 A box is being pushed and pulled across a rough surface by constant forces as shown in the diagram. The box is moving at a constant speed. By modelling the box as a particle, show that the magnitude of the resistance due to friction F is $25(3\sqrt{2} + 2\sqrt{3})$ N.
(4 marks)

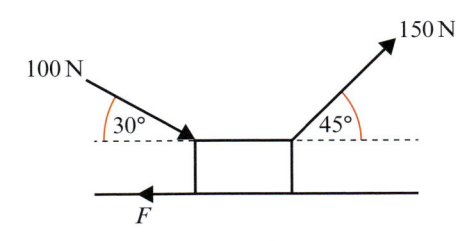

(E/P) 6 A trailer of mass 20 kg sits at rest on a rough horizontal plane. A force of 20 N acts on the trailer at an angle of 30° above the horizontal. Given that the trailer is in limiting equilibrium, work out the value of the coefficient of friction.
(6 marks)

7 A particle of mass 2 kg is moving down a rough plane that is inclined at α to the horizontal, where $\tan \alpha = \frac{3}{4}$. A force of P N acts horizontally upon the particle toward the plane. Given that the coefficient of friction is 0.3 and that the particle is moving at a constant velocity, calculate the value of P. **(7 marks)**

8 A particle of mass 0.5 kg is being pulled up a rough slope that is angled at 30° to the horizontal by a force of 5 N. The force acts at an angle of 30° above the slope. Given that the coefficient of friction is 0.1, calculate the acceleration of the particle. **(7 marks)**

9 A car of mass 2150 kg is travelling down a rough road that is inclined at 10° to the horizontal. The engine of the car applies a constant driving force of magnitude 700 N, which acts in the direction of travel of the car. The resistance to motion is modelled as a single constant force of magnitude F N that acts to oppose the motion of the car.

   a Given that the car is travelling in a straight line at a constant speed of 22 m s^{-1}, find the magnitude of F. **(3 marks)**

The driver brakes suddenly. In the subsequent motion the car continues to travel in a straight line, and the tyres skid along the road with the wheels locked, bringing the car to a standstill after 40 m. The driving force is removed, and the force due to resistance other than friction is again modelled as a constant force of magnitude F N.

   b Find the coefficient of friction between the tyres and the road. **(7 marks)**

   c Criticise this model with relation to:
     i the frictional forces acting on the car
     ii the motion of the car. **(2 marks)**

Challenge

A boat of mass 400 kg is being pulled up a rough slipway (a sloped road leading down into water) at a constant speed of 5 m s^{-1} by a winch (a type of pulley). The slipway is modelled as a plane inclined at an angle of 15° to the horizontal, and the boat is modelled as a particle. The coefficient of friction between the boat and the slipway is 0.2.

At the point when the boat is 8 m from the water-line, as measured along the line of greatest slope of the slipway, the winch cable snaps. Show that the boat will slide back down into the water, and calculate the total time from the winch cable breaking to the boat reaching the water-line.

Summary of key points

1 If a force is applied at an angle to the direction of motion, you can resolve it to find the component of the force that acts in the direction of motion.

2 The component of a force of magnitude F in a certain direction is $F \cos \theta$, where θ is the size of the angle between the force and the direction.

3 To solve problems involving inclined planes, it is usually easier to resolve forces parallel to and at right angles to the plane.

4 The maximum or limiting value of the friction between two surfaces, F_{MAX}, is given by $F_{MAX} = \mu F$ where μ is the coefficient of friction and R is the normal reaction between the two surfaces.

6 MOMENTUM AND IMPULSE

4.3

Learning objectives

After completing this chapter you should be able to:
- Calculate the momentum of a particle and the impulse of a force → pages 102–104
- Solve problems involving collisions using the principle of conservation of momentum → pages 104–109

Prior knowledge check

1. The forces $\mathbf{F}_1 = 3\mathbf{i} - 2\mathbf{j}$ and $\mathbf{F}_2 = 5\mathbf{i} + 4\mathbf{j}$ act on a particle. Find the magnitude and direction of the resultant force.
   ← Mechanics 1 Section 3.4

2. A particle moves in a straight line with constant acceleration.
   Given s = displacement in m, u = initial velocity in m s^{-1}, v = final velocity in m s^{-1}, a = acceleration in m s^{-2} and t = time in seconds, find:

   a v when $u = 3$, $a = 0.5$, $t = 5$

   b s when $u = 4.5$, $a = -1.5$, $t = 2$
   ← Mechanics 1 Sections 2.4, 2.5

3. A body of mass 2 kg is acted on by a force $\mathbf{F}$ N. The body starts from rest and moves in a straight line. After 5 seconds, the displacement of the body is 20 m. Find the magnitude of $\mathbf{F}$.
   ← Mechanics 1 Section 4.3

Newton's cradle demonstrates the **principle of conservation of momentum**. When the first ball collides with the second, the first ball stops, but its momentum is transferred to the second ball, then the third, then the fourth, until it reaches the very last ball.

6.1 Momentum in one dimension

You can calculate the momentum of a particle and the impulse of a force.

- **The momentum of a body of mass m which is moving with velocity v is mv.**

If m is in kg and v is in m s^{-1} then the units of momentum will be kg m s^{-1}. However, since kg m s^{-1} = (kg m s^{-2}) s and kg m s^{-2} are the units of force ($F = ma$) you can also measure momentum in newton seconds (N s).

Notation Velocity is a vector quantity and mass is a scalar, so momentum is a vector quantity.

Example 1 SKILLS PROBLEM-SOLVING

Find the magnitude of the momentum of:
a a cricket ball of mass 400 g moving at 18 m s^{-1}
b a lorry of mass 5 tonnes moving at 0.3 m s^{-1}.

a Momentum = mass × velocity
   Magnitude of momentum = $\frac{400}{1000} \times 18$
   = 7.2 N s

— The mass must be in kg, so divide by 1000.
— The units can be N s or kg m s^{-1}.

b Momentum = mass × velocity
   Magnitude of momentum = (5 × 1000) × 0.3
   = 1500 kg m s^{-1}

— 1 tonne = 1000 kg.

- **If a constant force F acts for time t then we define the impulse of the force to be Ft.**

If F is in N and t is in s then the units of impulse will be N s.

Notation Force is a vector quantity and time is a scalar, so impulse is a vector quantity.

Examples of an impulse include a bat hitting a ball, a snooker ball hitting another ball or a jerk in a string when it suddenly goes tight. In all these cases the time for which the force acts is very small but the force is quite large. This means that the product of the two, which gives the impulse, is of reasonable size. However, there is no theoretical limit on the size of t.

Suppose a body of mass m is moving with an initial speed u and is then acted upon by a force F for time t. This results in its final speed being v.

Its acceleration is given by $a = \frac{v-u}{t}$

Substituting into $F = ma$: $F = m\left(\frac{v-u}{t}\right)$

$Ft = m(v - u)$ — The impulse I of the force is given by $I = Ft$.
   $= mv - mu$

- $I = mv - mu$
  Impulse = final momentum − initial momentum
  Impulse = change in momentum

This is called the **impulse–momentum principle**.

Watch out This is a vector equation, so for motion in a straight line a positive direction must be chosen and each value must be given the correct sign.

MOMENTUM AND IMPULSE — CHAPTER 6

Example 2 — SKILLS: PROBLEM-SOLVING

A body of mass 2 kg is initially at rest on a smooth horizontal plane. A horizontal force of magnitude 4.5 N acts on the body for 6 s. Find:

a the magnitude of the impulse given to the body by the force
b the final speed of the body.

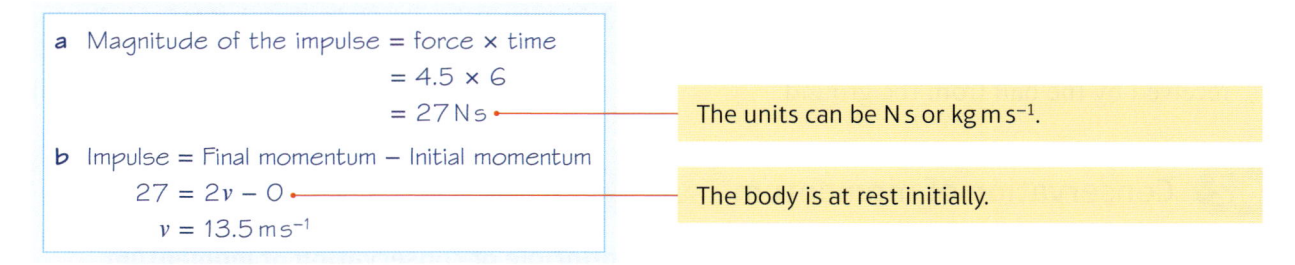

a Magnitude of the impulse = force × time
$= 4.5 \times 6$
$= 27 \text{ N s}$ — The units can be N s or kg m s^{-1}.

b Impulse = Final momentum − Initial momentum
$27 = 2v - 0$ — The body is at rest initially.
$v = 13.5 \text{ m s}^{-1}$

Example 3 — SKILLS: PROBLEM-SOLVING

A ball of mass 0.2 kg hits a fixed vertical wall at right angles with speed 3.5 m s^{-1}. The ball rebounds with speed 2.5 m s^{-1}. Find the magnitude of the impulse exerted on the wall by the ball.

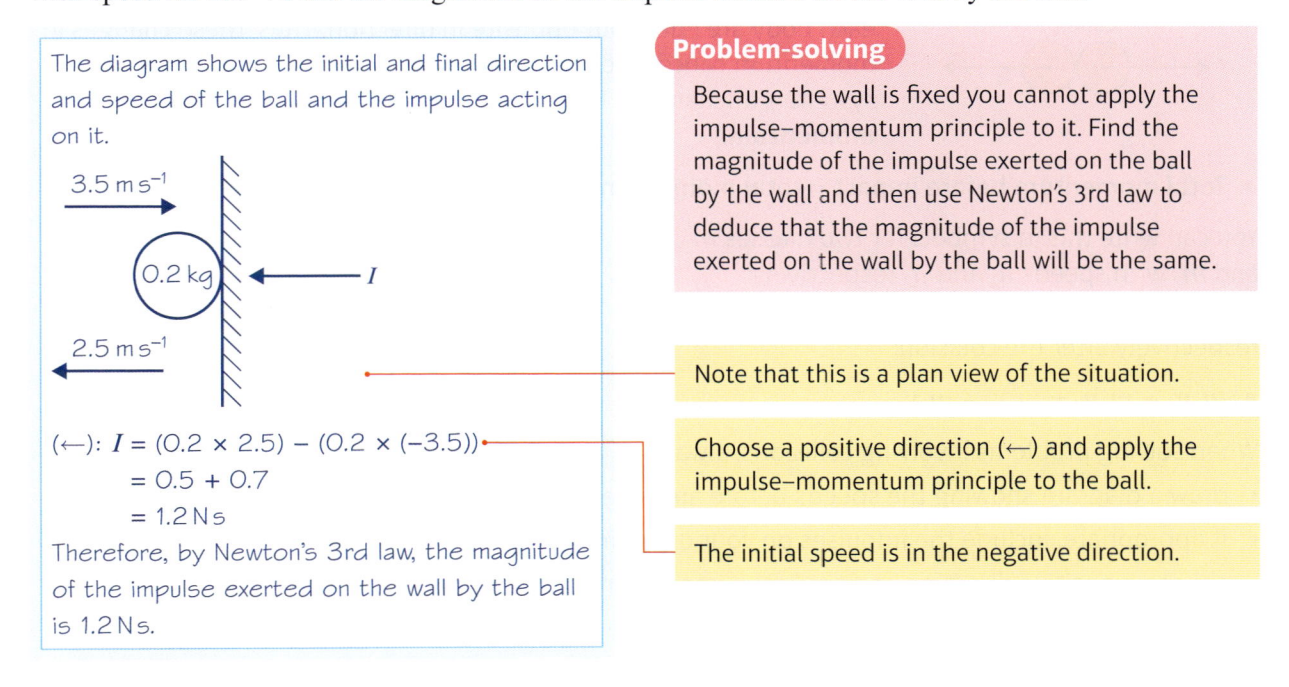

The diagram shows the initial and final direction and speed of the ball and the impulse acting on it.

Problem-solving
Because the wall is fixed you cannot apply the impulse–momentum principle to it. Find the magnitude of the impulse exerted on the ball by the wall and then use Newton's 3rd law to deduce that the magnitude of the impulse exerted on the wall by the ball will be the same.

Note that this is a plan view of the situation.

$(\leftarrow): I = (0.2 \times 2.5) - (0.2 \times (-3.5))$ — Choose a positive direction ($\leftarrow$) and apply the impulse–momentum principle to the ball.
$= 0.5 + 0.7$
$= 1.2 \text{ N s}$ — The initial speed is in the negative direction.

Therefore, by Newton's 3rd law, the magnitude of the impulse exerted on the wall by the ball is 1.2 N s.

Exercise 6A — SKILLS: PROBLEM-SOLVING

1 A ball of mass 0.5 kg is at rest when it is struck by a bat and receives an impulse of 15 N s. Find its speed immediately after it is struck.

2 A ball of mass 0.3 kg moving along a horizontal surface hits a fixed vertical wall at right angles with speed 3.5 m s^{-1}. The ball rebounds at right angles to the wall. Given that the magnitude of the impulse exerted on the ball by the wall is 1.8 N s, find the speed of the ball just after it rebounds.

3 A toy car of mass 0.2 kg is pushed from rest along a smooth horizontal floor by a horizontal force of magnitude 0.4 N for 1.5 s. Find its speed at the end of the 1.5 s.

(E) 4 A ball of mass 0.2 kg, moving along a horizontal surface, hits a fixed vertical wall at right angles. The ball rebounds at right angles to the wall with speed $3.5\,\mathrm{m\,s^{-1}}$. Given that the magnitude of the impulse exerted on the ball by the wall is 2 N s, find the speed of the ball just before it hits the wall. **(3 marks)**

(E/P) 5 A ball of mass 0.2 kg is dropped from a height of 2.5 m above horizontal ground. After hitting the ground it bounces to a height of 1.8 m above the ground. Find the magnitude of the impulse received by the ball from the ground. **(4 marks)**

6.2 Conservation of momentum

You can solve problems involving collisions using the **principle of conservation of momentum**.

By Newton's 3rd law, when two bodies collide, each one exerts an equal and opposite force on the other. They are in contact for the same time, so they each exert an impulse on the other of equal magnitude but in opposite directions.

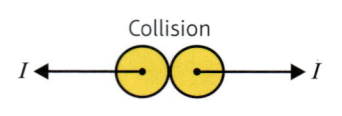

By the impulse–momentum principle, the changes in momentum of each body are equal but opposite in direction. Thus, these changes in momentum cancel each other out, and the momentum of the whole system is unchanged. This is called the **principle of conservation of momentum**.

- **Total momentum before impact = total momentum after impact**

You can write this in symbols for two masses m_1 and m_2 with speeds u_1 and u_2 respectively before the collision, and speeds v_1 and v_2 respectively after the collision:

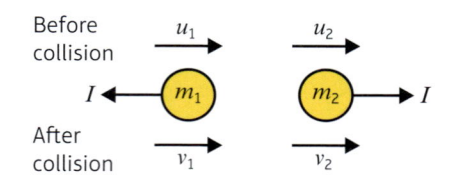

- $m_1 u_1 + m_2 u_2 = m_1 v_1 + m_2 v_2$

When solving problems involving collisions, always:

- draw a diagram showing the speeds before and after the collision with arrows
- if appropriate, include the impulses on your diagram with arrows
- choose a positive direction and apply the impulse–momentum principle and/or the principle of conservation of momentum.

Example 4   SKILLS   PROBLEM-SOLVING

A particle P of mass 2 kg is moving with speed $3\,\mathrm{m\,s^{-1}}$ on a smooth horizontal plane. Particle Q of mass 3 kg is at rest on the plane. Particle P collides with particle Q and after the collision Q moves off with speed $\frac{7}{3}\,\mathrm{m\,s^{-1}}$. Find:

a the speed and direction of motion of P after the collision

b the magnitude of the impulse received by P in the collision.

Online Explore particle collisions using GeoGebra.

a

P 3 m s⁻¹ → ... Q 0 m s⁻¹ →
I ← (2 kg) (3 kg) → I
v m s⁻¹ → ... $\frac{7}{3}$ m s⁻¹ →

Conservation of momentum: (→)

$m_1 u_1 + m_2 u_2 = m_1 v_1 + m_2 v_2$

$(2 \times 3) + (3 \times 0) = 2v + \left(3 \times \frac{7}{3}\right)$

$6 = 2v + 7$

$v = -\frac{1}{2}$

The direction of motion of P is reversed by the collision and its speed is $\frac{1}{2}$ m s⁻¹.

b For Q: (→) $I = 3\left(\frac{7}{3} - 0\right)$

$= 7$ N s

Alternatively, for P: (←)

$I = 2((-v) - (-3))$

$= 2\left(\frac{1}{2} + 3\right)$

$= 7$ N s

So the impulse received by P has magnitude 7 N s.

- Draw a diagram showing the speeds before and after the collision (with arrows) and the impulses (with arrows).
- Choose a positive direction and apply the principle of conservation of momentum.
- Since v is negative, P moves in the opposite direction after the collision.
- **Watch out** The direction of motion of P in your answer must be with reference to the original direction of motion of P. Do not use the words left or right.
- To find the impulse, consider one particle and apply the impulse–momentum principle. Here it is easier to consider Q.
- Since each particle receives an impulse of equal magnitude, the magnitude of the impulse received by P is also 7 N s.

Example 5   SKILLS   PROBLEM-SOLVING

Two particles A and B, of masses 2 kg and 4 kg respectively, are moving toward each other in opposite directions along the same straight line on a smooth horizontal surface. The particles collide. Before the collision, the speeds of A and B are 3 m s⁻¹ and 2 m s⁻¹ respectively. After the collision, the direction of motion of A is reversed and its speed is 2 m s⁻¹. Find:

a the speed and direction of B after the collision

b the magnitude of the impulse given by A to B in the collision.

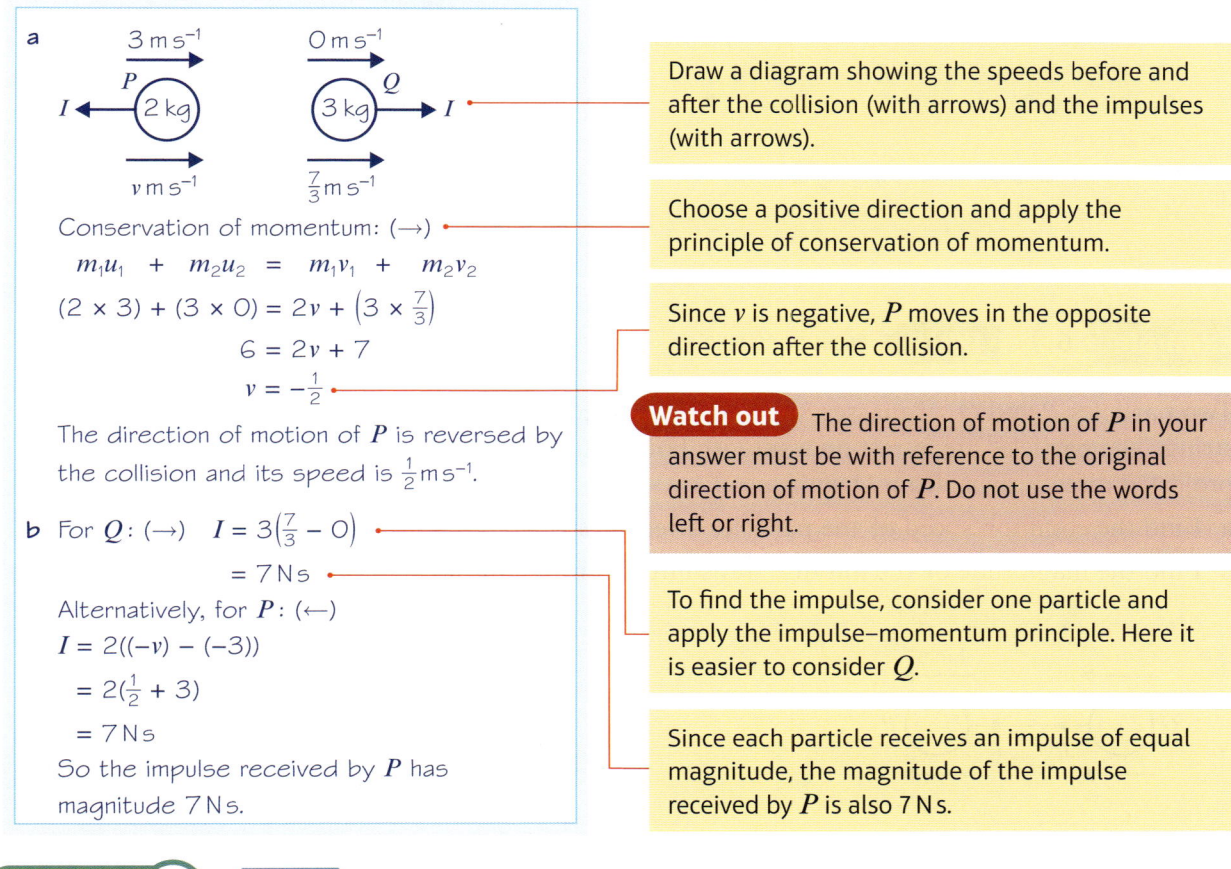

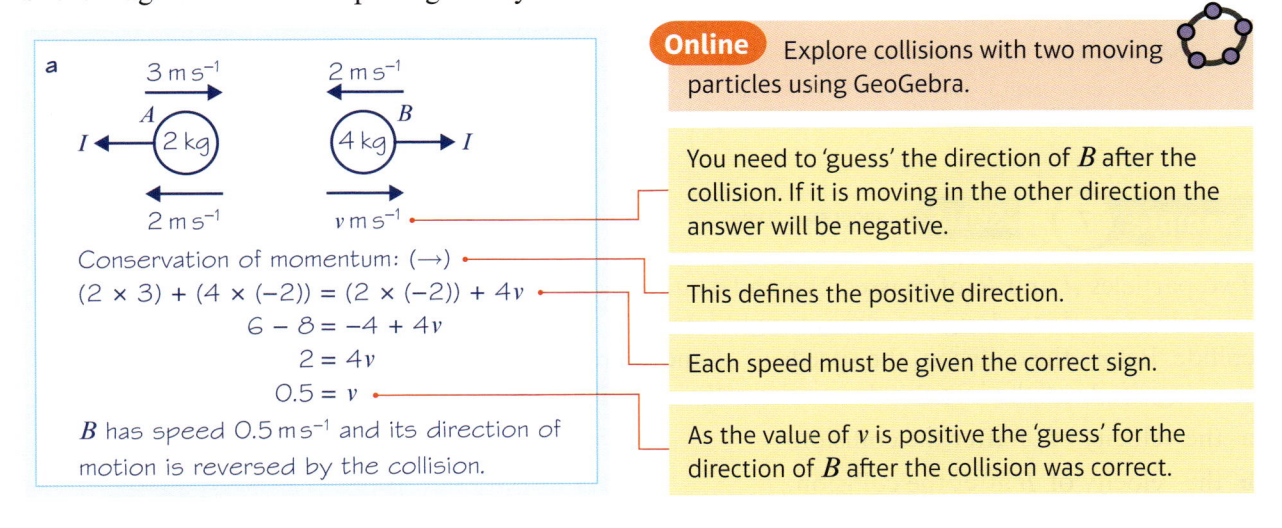

Online Explore collisions with two moving particles using GeoGebra.

a

A 3 m s⁻¹ → ... B ← 2 m s⁻¹
I ← (2 kg) (4 kg) → I
← 2 m s⁻¹ ... v m s⁻¹ →

Conservation of momentum: (→)

$(2 \times 3) + (4 \times (-2)) = (2 \times (-2)) + 4v$

$6 - 8 = -4 + 4v$

$2 = 4v$

$0.5 = v$

B has speed 0.5 m s⁻¹ and its direction of motion is reversed by the collision.

- You need to 'guess' the direction of B after the collision. If it is moving in the other direction the answer will be negative.
- This defines the positive direction.
- Each speed must be given the correct sign.
- As the value of v is positive the 'guess' for the direction of B after the collision was correct.

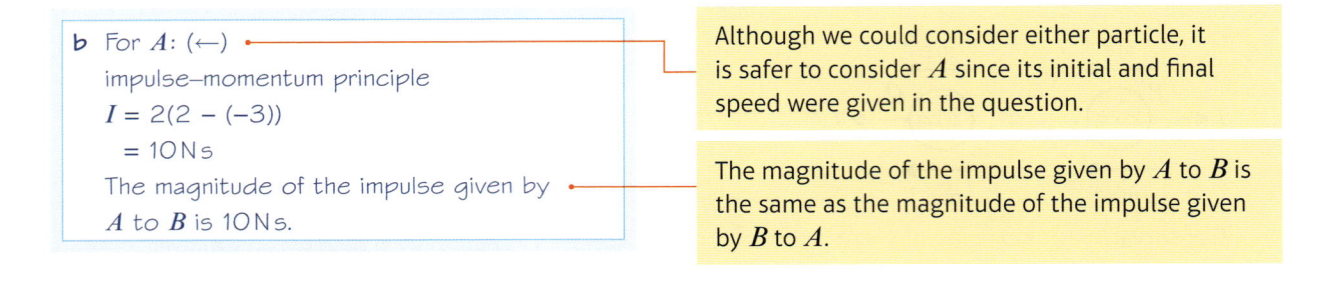

b For A: ($\leftarrow$)
impulse–momentum principle
$I = 2(2 - (-3))$
$\quad = 10\,\text{N s}$
The magnitude of the impulse given by A to B is $10\,\text{N s}$.

Although we could consider either particle, it is safer to consider A since its initial and final speed were given in the question.

The magnitude of the impulse given by A to B is the same as the magnitude of the impulse given by B to A.

Example 6   SKILLS   PROBLEM-SOLVING

Two particles P and Q, of masses $8\,\text{kg}$ and $2\,\text{kg}$ respectively, are connected by a light inextensible string. The particles are at rest on a smooth horizontal plane with the string slack. Particle P is projected directly away from Q with speed $4\,\text{m s}^{-1}$.

a Find the common speed of the particles after the string goes taut.

b Find the magnitude of the impulse transmitted through the string when it goes taut.

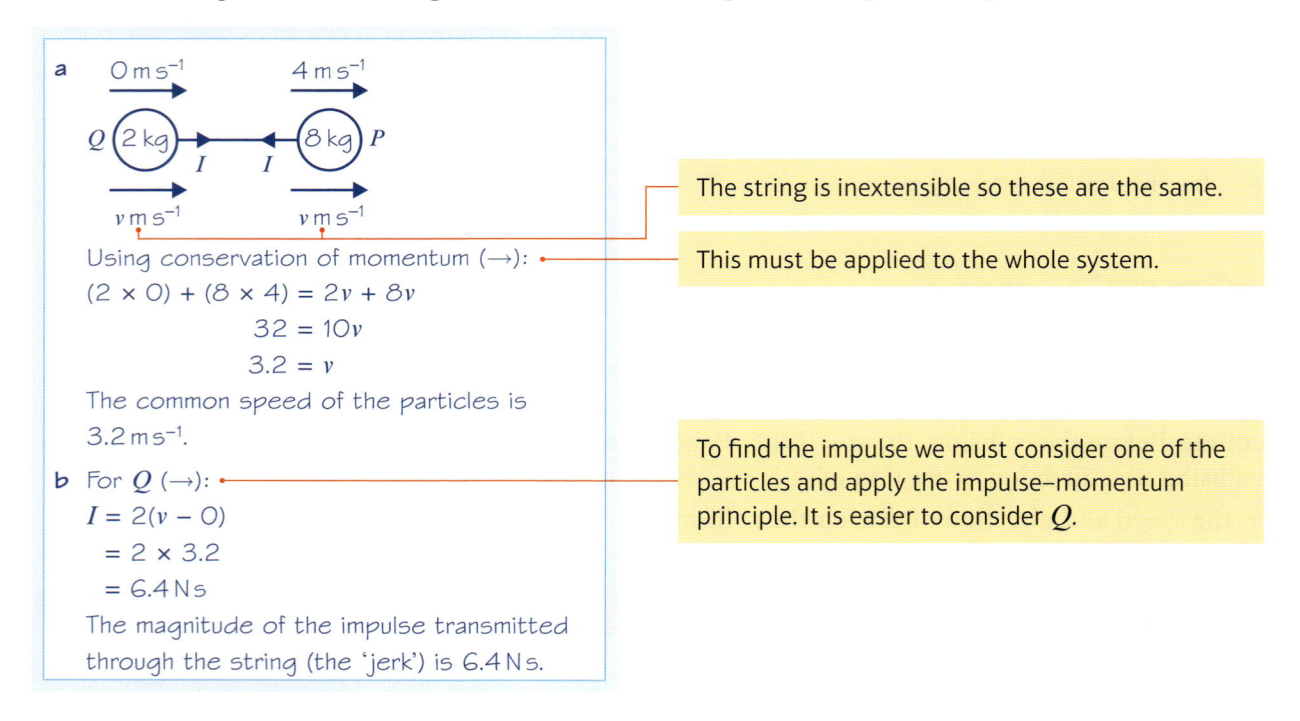

a [Diagram: Q (2 kg) at $0\,\text{m s}^{-1}$, P (8 kg) at $4\,\text{m s}^{-1}$, both moving at $v\,\text{m s}^{-1}$ after, impulse I between them]

Using conservation of momentum ($\rightarrow$):
$(2 \times 0) + (8 \times 4) = 2v + 8v$
$32 = 10v$
$3.2 = v$
The common speed of the particles is $3.2\,\text{m s}^{-1}$.

b For Q ($\rightarrow$):
$I = 2(v - 0)$
$\quad = 2 \times 3.2$
$\quad = 6.4\,\text{N s}$
The magnitude of the impulse transmitted through the string (the 'jerk') is $6.4\,\text{N s}$.

The string is inextensible so these are the same.

This must be applied to the whole system.

To find the impulse we must consider one of the particles and apply the impulse–momentum principle. It is easier to consider Q.

Example 7   SKILLS   PROBLEM-SOLVING

Two particles A and B, of masses $2\,\text{kg}$ and $4\,\text{kg}$ respectively, are moving toward each other in opposite directions along the same straight line on a smooth horizontal surface. The particles collide. Before the collision, the speeds of A and B are $3\,\text{m s}^{-1}$ and $2\,\text{m s}^{-1}$ respectively. Given that the magnitude of the impulse due to the collision is $7\,\text{N s}$, find:

a the velocity of A after the collision

b the velocity of B after the collision.

MOMENTUM AND IMPULSE   CHAPTER 6

```
         3 m s⁻¹        2 m s⁻¹
           →              ←
          A              B
  7 N s ← (2 kg)       (4 kg) → 7 N s
           →              →
         v₁ m s⁻¹       v₂ m s⁻¹
```

For A: $(\leftarrow)$
$$7 = 2(v_1 - (-3))$$
$$7 = 2(v_1 + 3)$$
$$3.5 = v_1 + 3$$
$$0.5 = v_1$$

For B: $(\rightarrow)$
$$7 = 4(v_2 - (-2))$$
$$1.75 = v_2 + 2$$
$$-0.25 = v_2$$

a The direction of motion of A is reversed and its speed is $0.5\,\text{m s}^{-1}$.

b The direction of motion of B is unchanged and its speed is $0.25\,\text{m s}^{-1}$.

Online Explore particle collisions with known impulse using GeoGebra.

The diagram should show the speeds and impulses with arrows.

Again you need to guess which way the particles go after the collision. There are three sensible possibilities: the one shown, $\overset{\leftarrow}{v_1}\,\overset{\rightarrow}{v_2}$ or $\overset{\leftarrow}{v_1}\,\overset{\leftarrow}{v_2}$.

To find either v_1 or v_2, consider one particle only, choose a positive direction and apply the impulse–momentum principle.

As v_1 is positive, the guess for the direction of A was correct.

As v_2 is negative, the guess for the direction of B was incorrect and it travels in the opposite direction to that shown on the diagram.

Watch out The question asks for the **velocities** so you need to state the speed and the direction of motion.

Exercise 6B SKILLS PROBLEM-SOLVING

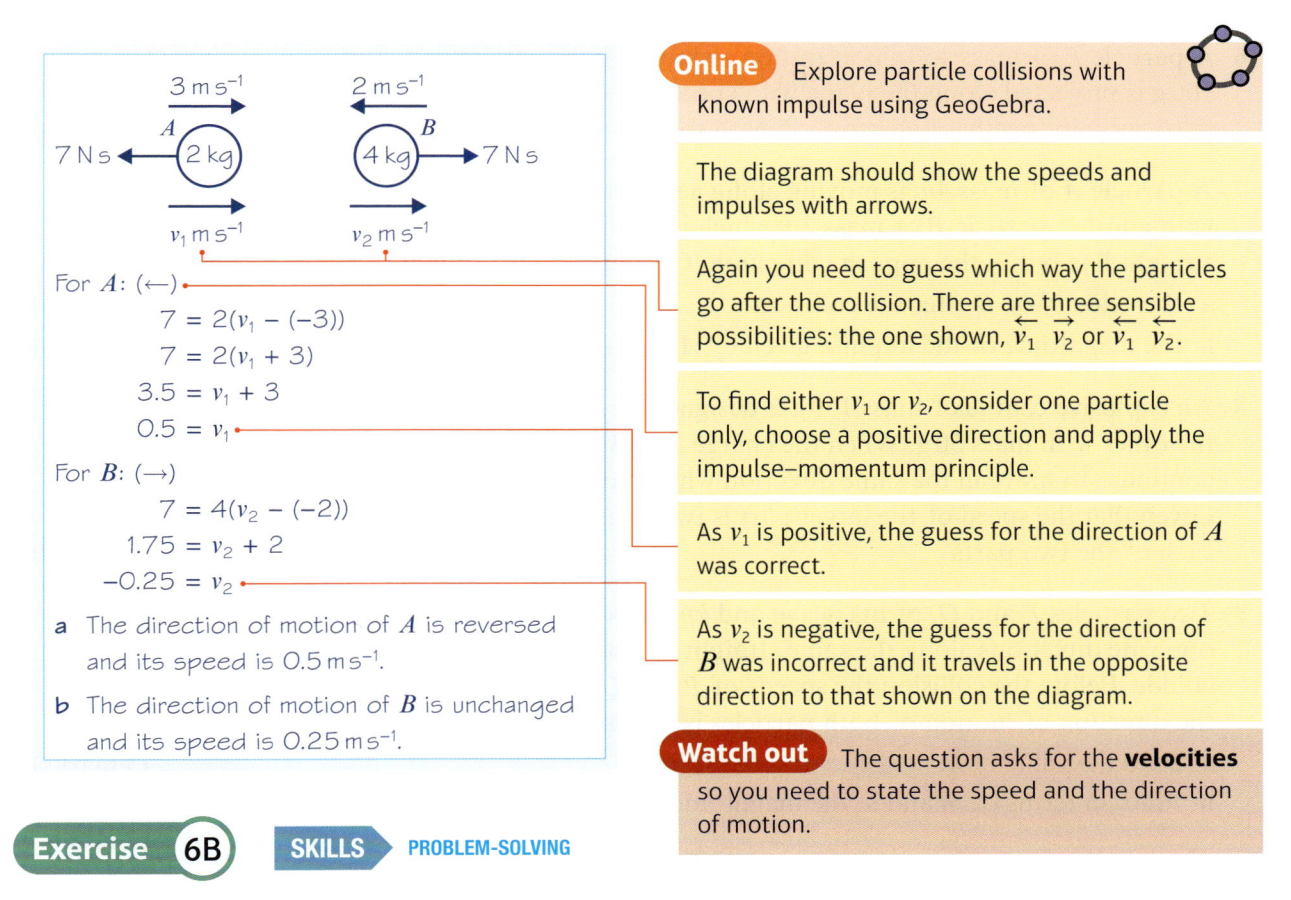

1. A particle P of mass 2 kg is moving on a smooth horizontal plane with speed $4\,\text{m s}^{-1}$. It collides with a second particle Q of mass 1 kg which is at rest. After the collision P has speed $2\,\text{m s}^{-1}$ and it continues to move in the same direction. Find the speed of Q after the collision.

2. A railway truck of mass 25 tonnes moving at $4\,\text{m s}^{-1}$ collides with a stationary truck of mass 20 tonnes. As a result of the collision the trucks couple together. Find the common speed of the trucks after the collision.

3. Particles A and B have masses 0.5 kg and 0.2 kg respectively. They are moving with speeds $5\,\text{m s}^{-1}$ and $2\,\text{m s}^{-1}$ respectively in the same direction along the same straight line on a smooth horizontal surface when they collide. After the collision, A continues to move in the same direction with speed $4\,\text{m s}^{-1}$. Find the speed of B after the collision.

4. A particle of mass 2 kg is moving on a smooth horizontal plane with speed $4\,\text{m s}^{-1}$. It collides with a second particle of mass 1 kg which is at rest. After the collision the particles join together.
   a Find the common speed of the particles after the collision.
   b Find the magnitude of the impulse in the collision.

(E) 5. Two particles A and B, of masses 2 kg and 5 kg respectively, are moving toward each other along the same straight line on a smooth horizontal surface. The particles collide. Before the collision the speeds of A and B are $6\,\text{m s}^{-1}$ and $4\,\text{m s}^{-1}$ respectively. After the collision the direction of motion of A is reversed and its speed is $1.5\,\text{m s}^{-1}$. Find:
   a the speed and direction of B after the collision (3 marks)
   b the magnitude of the impulse given by A to B in the collision. (3 marks)

6 A particle P of mass 150 g is at rest on a smooth horizontal plane. A second particle Q of mass 100 g is projected along the plane with speed u m s^{-1} and collides directly with P. On impact the particles join together and move on with speed 4 m s^{-1}. Find the value of u. **(4 marks)**

7 A particle A of mass $4m$ is moving along a smooth horizontal surface with speed $2u$. It collides with another particle B of mass $3m$ which is moving with the same speed along the same straight line but in the opposite direction. Given that A is brought to rest by the collision, find:
  a the velocity of B after the collision **(3 marks)**
  b the magnitude of the impulse given by A to B in the collision. **(3 marks)**

8 An explosive charge of mass 150 g is designed to split into two parts, one with mass 100 g and the other with mass 50 g. When the charge is moving at 4 m s^{-1} it splits and the larger part continues to move in the same direction while the smaller part moves in the opposite direction. Given that the speed of the larger part is twice the speed of the smaller part, find the speeds of each of the two parts. **(3 marks)**

9 Two particles P and Q, of masses m and km respectively, are moving toward each other in opposite directions along the same straight line on a smooth horizontal surface. The particles collide. Before the collision the speeds of P and Q are $3u$ and u respectively. After the collision the direction of motion of both particles is reversed and the speed of each particle is halved.
  a Find the value of k. **(4 marks)**
  b Find, in terms of m and u, the magnitude of the impulse given by P to Q in the collision. **(3 marks)**

10 Two particles A and B, of masses 4 kg and 2 kg respectively, are connected by a light inextensible string. The particles are at rest on a smooth horizontal plane with the string slack. Particle A is projected directly away from B with speed u m s^{-1}. When the string goes taut the impulse transmitted through the string has magnitude 6 N s. Find:
  a the common speed of the particles just after the string goes taut **(4 marks)**
  b the value of u. **(3 marks)**

11 Two particles P and Q, of masses 3 kg and 2 kg respectively, are moving along the same straight line on a smooth horizontal surface. The particles collide. After the collision both the particles are moving in the same direction, the speed of P is 1 m s^{-1} and the speed of Q is 1.5 m s^{-1}. The magnitude of the impulse of P on Q is 9 N s. Find:
  a the speed and direction of P before the collision **(3 marks)**
  b the speed and direction of Q before the collision. **(3 marks)**

12 Two particles A and B are moving in the same direction along the same straight line on a smooth horizontal surface. The particles collide. Before the collision the speed of B is 1.5 m s^{-1}. After the collision the direction of motion of both particles is unchanged, the speed of A is 2.5 m s^{-1} and the speed of B is 3 m s^{-1}.
  a Given that the mass of A is three times the mass of B, find the speed of A before the collision. **(4 marks)**
  b Given that the magnitude of the impulse on A in the collision is 3 N s, find the mass of A. **(3 marks)**

MOMENTUM AND IMPULSE   CHAPTER 6

Challenge

Particle P has mass $3m$ kg and particle Q has mass m kg. The particles are moving in opposite directions along the same straight line on a smooth horizontal plane when they collide directly. Immediately before the collision, the speed of P is u_1 m s^{-1} and the speed of Q is u_2 m s^{-1}. In the collision, the direction of motion of P is unchanged and the direction of Q is reversed. Immediately after the collision, the speed of P is $\frac{1}{4}u_1$ and the speed of Q is $\frac{1}{2}u_2$. Show that $u_1 = \frac{2}{3}u_2$.

Chapter review 6

1 A particle P of mass $3m$ is moving along a straight line with constant speed $2u$. It collides with another particle Q of mass $4m$ which is moving with speed u along the same line but in the opposite direction. As a result of the collision P is brought to rest.
  a Find the speed of Q after the collision and state its direction of motion.
  b Find the magnitude of the impulse exerted by Q on P in the collision.

2 A pile driver of mass 1000 kg drives a pile of mass 200 kg vertically into the ground. The driver falls freely a vertical distance of 10 m before hitting the pile. Immediately after the driver impacts with the pile it can be assumed that they both move with the same velocity. By modelling the pile and the driver as particles, find:
  a the speed of the driver immediately before it hits the pile   **(2 marks)**
  b the common speed of the pile and driver immediately after the impact.   **(3 marks)**
The ground provides a constant resistance to the motion of the pile driver of magnitude 120 000 N.
  c Find the distance that the pile is driven into the ground before coming to rest.   **(2 marks)**
  d Comment on this model in relation to the motion of the pile and driver immediately after impact.   **(1 mark)**

3 A car of mass 800 kg is travelling along a straight horizontal road. A constant retarding force of F N reduces the speed of the car from 18 m s^{-1} to 12 m s^{-1} in 2.4 s. Calculate:
  a the value of F **(4 marks)**
  b the distance moved by the car in these 2.4 s.   **(3 marks)**

4 Two particles A and B, of masses 0.2 kg and 0.3 kg respectively, are free to move in a smooth horizontal groove. Initially B is at rest and A is moving toward B with a speed of 4 m s^{-1}. After the impact the speed of B is 1.5 m s^{-1}. Find:
  a the speed of A after the impact   **(3 marks)**
  b the magnitude of the impulse of B on A during the impact.   **(3 marks)**

(E) 5 A railway truck P of mass 2000 kg is moving along a straight horizontal track with speed $10\,\text{m}\,\text{s}^{-1}$. The truck P collides with a truck Q of mass 3000 kg, which is at rest on the same track. Immediately after the collision Q moves with speed $5\,\text{m}\,\text{s}^{-1}$. Calculate:

   a the speed of P immediately after the collision   **(3 marks)**

   b the magnitude of the impulse exerted by P on Q during the collision.   **(3 marks)**

(E) 6 A particle P of mass 1.5 kg is moving along a straight horizontal line with speed $3\,\text{m}\,\text{s}^{-1}$. Another particle Q of mass 2.5 kg is moving, in the opposite direction, along the same straight line with speed $4\,\text{m}\,\text{s}^{-1}$. The particles collide. Immediately after the collision the direction of motion of P is reversed and its speed is $2.5\,\text{m}\,\text{s}^{-1}$.

   a Calculate the speed of Q immediately after the impact.   **(3 marks)**

   b State whether or not the direction of motion of Q is changed by the collision.   **(1 mark)**

   c Calculate the magnitude of the impulse exerted by Q on P, giving the units of your answer.   **(3 marks)**

(E/P) 7 A particle A of mass m is moving with speed $2u$ in a straight line on a smooth horizontal table. It collides with another particle B of mass km which is moving in the same straight line on the table with speed u in the opposite direction to A. In the collision, the particles form a single particle which moves with speed $\frac{2}{3}u$ in the original direction of A's motion.

   Find the value of k.   **(3 marks)**

(E/P) 8 A metal pin of mass 2 kg is driven vertically into the ground by a blow from a sledgehammer of mass 10 kg. The hammer falls vertically onto the pin, its speed just before impact being $9\,\text{m}\,\text{s}^{-1}$. In a model of the situation it is assumed that, after impact, the pin and the hammer stay in contact and move together before coming to rest.

   a Find the speed of the pin immediately after impact.   **(3 marks)**

   The pin moves 3 cm into the ground before coming to rest. Assuming in this model that the ground exerts a constant resistive force of magnitude R newtons as the pin is driven down,

   b find the value of R.   **(5 marks)**

   c State one way in which this model might be refined to be more realistic.   **(1 mark)**

MOMENTUM AND IMPLUSE — CHAPTER 6

Summary of key points

1. The **momentum** of a body of mass m which is moving with speed v is mv.
   The units of momentum can be N s or kg m s^{-1}.

2. If a constant force F acts for a time t then we define the **impulse** of the force to be Ft.
   The units of impulse are N s.

3. The **impulse–momentum principle**:
   Impulse = final momentum − initial momentum
   Impulse = change in momentum
   $$I = mv - mu$$
   where m is the mass of the body, u is the initial speed and v is the final speed.

4. **Principle of conservation of momentum**:
   Total momentum before impact = total momentum after impact

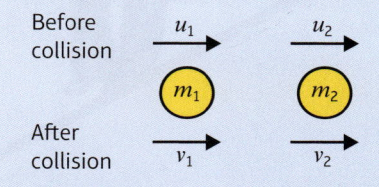

   Before collision: $u_1 \rightarrow$ m_1, $u_2 \rightarrow$ m_2
   After collision: $v_1 \rightarrow$, $v_2 \rightarrow$

   $$m_1 u_1 + m_2 u_2 = m_1 v_1 + m_2 v_2$$

   where a body of mass m_1 moving with speed u_1 collides with a body of mass m_2 moving with speed u_2, v_1 and v_2 are the speeds of m_1 and m_2 after the collision respectively.

7 STATICS OF A PARTICLE

5.1
5.2
5.3

Learning objectives

After completing this chapter you should be able to:
- Find an unknown force when a system is in equilibrium → pages 113–116
- Solve statics problems involving weight, tension and pulleys → pages 117–121
- Understand and solve problems involving limiting equilibrium → pages 121–126

Prior knowledge check

1. A particle of mass 2 kg sits on a rough plane that is inclined at 45° to the horizontal. A force of 10 N acts parallel to and up the plane. Given that the particle is on the point of moving, work out the coefficient of friction μ. ← **Mechanics 1 Section 5.2**

2. A particle of negligible mass has three forces acting upon it as shown in the diagram below. Work out the magnitude and direction of the resultant force relative to the horizontal dashed line.

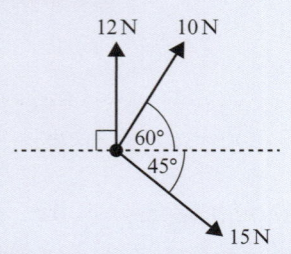

   12 N 10 N
   60°
   45°
   15 N
   ← **Mechanics 1 Section 5.1**

A tightrope walker uses a mathematical model to calculate the tension in his wire. This allows him to make sure that the wire is strong enough to hold his weight safely.

STATICS OF A PARTICLE   CHAPTER 7

7.1 Static particles

- A particle or rigid body is in **static** equilibrium if it is at rest and the resultant force acting on the particle or rigid body is zero.

To solve problems in statics you should:
- Draw a diagram showing clearly the forces acting on the particle.
- Resolve the forces into horizontal and vertical components or, if the particle is on an inclined plane, into components parallel and perpendicular to the plane.
- Set the sum of the components in each direction equal to zero.
- Solve the resulting equations to find the unknown force(s).

Example 1   SKILLS   PROBLEM-SOLVING

The diagram shows a particle in equilibrium under the forces shown. By resolving horizontally and vertically, find the magnitudes of the forces P and Q.

Hint The particle is not accelerating, so $a = 0$. $F = ma = 0$

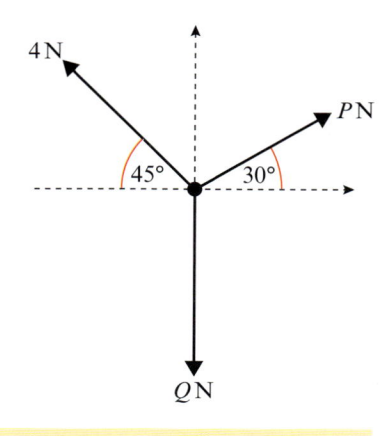

Method 1:
$R(\rightarrow)$, $P\cos 30° - 4\cos 45° = 0$
$R(\uparrow)$, $P\sin 30° + 4\sin 45° - Q = 0$

$P = \dfrac{4\cos 45°}{\cos 30°} = 3.27$ (3 s.f.)

$Q = P\sin 30° + 4\sin 45° = 4.46$ (3 s.f.)

Method 2:

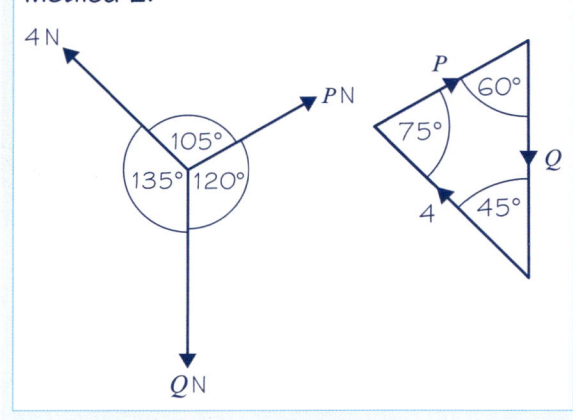

Resolve horizontally and vertically. Equate the sum of the forces to zero as there is no acceleration (the particle is in equilibrium).

Solve the first equation to find P (as there is only one unknown quantity), and then use your value for P in the second equation to find Q.

If exact answers are required these would be

$P = \dfrac{4\sqrt{6}}{3}$ and $Q = \dfrac{2(\sqrt{6} + 3\sqrt{2})}{3}$

Problem-solving

You can use a vector diagram to solve equilibrium problems involving three forces. Because the particle is in equilibrium, the three forces will form a **closed triangle**.

If the angle between forces on the force diagram is θ, the angle between those forces on the **triangle of forces** is $180° - \theta$.

The length of each side of the triangle is the magnitude of the force.

So $\dfrac{4}{\sin 60°} = \dfrac{P}{\sin 45°} = \dfrac{Q}{\sin 75°}$

$P = \dfrac{4 \sin 45°}{\sin 60°} = \dfrac{4\sqrt{6}}{3}$ N

$Q = \dfrac{4 \sin 75°}{\sin 60°} = \dfrac{2(\sqrt{6} + 3\sqrt{2})}{3}$ N

Use the sine rule. ← **Pure Maths 1 Section 6.2**

Watch out This method works only for a particle in equilibrium. If the resultant force is not zero, the vector diagram will not be a closed triangle.

Example 2 — SKILLS: PROBLEM-SOLVING

The diagram shows a particle in equilibrium on an inclined plane under the forces shown. Find the magnitude of the force P and the size of the angle α.

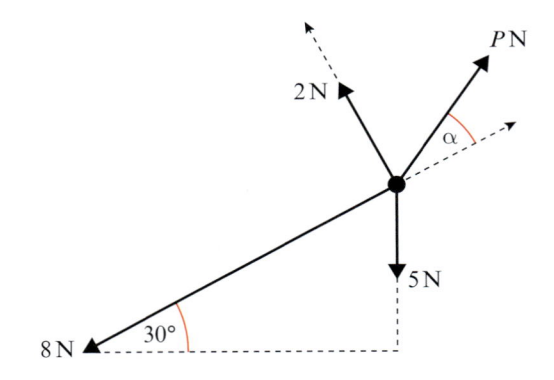

$R(\nearrow)$, $P \cos \alpha - 8 - 5 \sin 30° = 0$
∴ $P \cos \alpha = 8 + 5 \sin 30°$ (1)

$R(\nwarrow)$, $P \sin \alpha + 2 - 5 \cos 30° = 0$
∴ $P \sin \alpha = 5 \cos 30° - 2$ (2)

Divide equation (2) by equation (1) to give:

$\tan \alpha = \dfrac{5 \cos 30° - 2}{8 + 5 \sin 30°} = \dfrac{2.330}{10.5} = 0.222$

∴ $\alpha = 12.5°$ (3 s.f.)

Substitute into equation (1):
$P \sin 12.5...° = 5 \cos 30° - 2$
∴ $P = 10.8$ (3 s.f.)

Resolve parallel to the plane. Take the direction up the plane as positive.

Rearrange the equation to make $P \cos \alpha$ the subject.

Resolve perpendicular to the plane. Rearrange the second equation to make $P \sin \alpha$ the subject.

After division use $\dfrac{\sin \alpha}{\cos \alpha} \equiv \tan \alpha$.

Use $\tan^{-1}$ and give your answer to three significant figures.

You could check your answers by substituting into equation (2).

Exercise 7A — SKILLS: PROBLEM-SOLVING

1 Each of the following diagrams shows a particle in static equilibrium. For each particle:
  i resolve the components in the x-direction
  ii resolve the components in the y-direction
  iii find the magnitude of any unknown forces (marked P and Q) and the size of any unknown angles (marked θ).

STATICS OF A PARTICLE CHAPTER 7 115

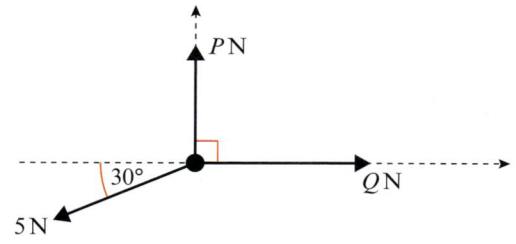

a, b, 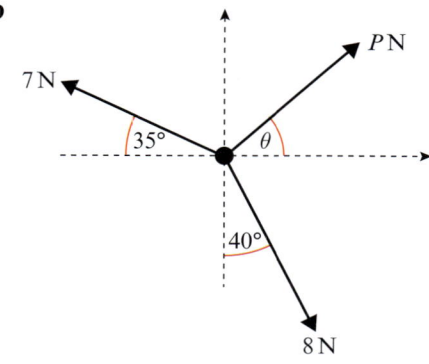 c, 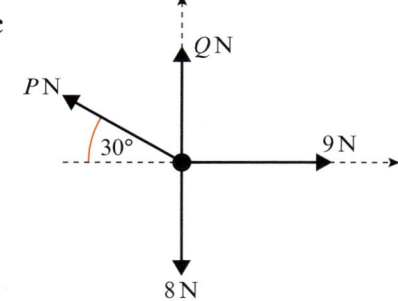 d, 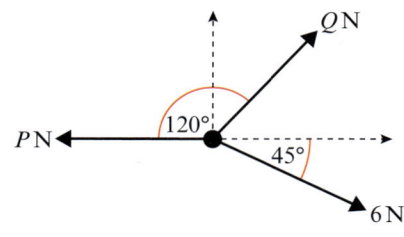 e, 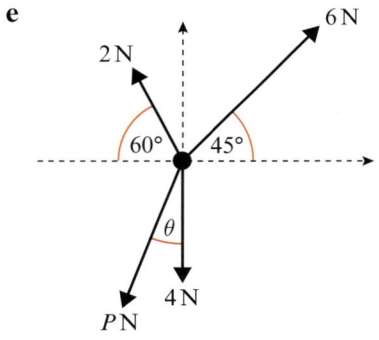 f (force diagrams)

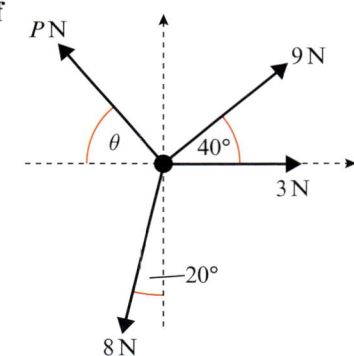

2 For each of the following particles in static equilibrium:
   i draw a triangle of forces diagram
   ii use trigonometry to find the magnitude of any unknown forces (marked P and Q) and the size of any unknown angles (marked θ).

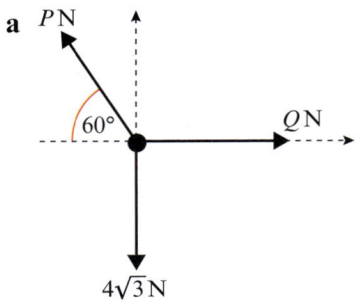

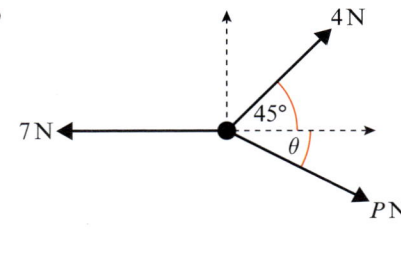

Hint The triangle of forces diagram for part **a** is:

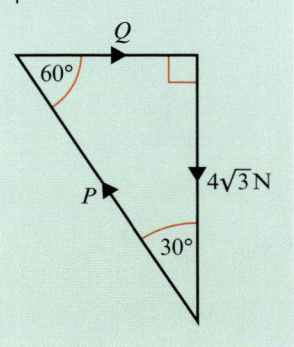

3 Each of these particles rests in equilibrium on an inclined plane under the forces shown. In each case, find the magnitude of forces P and Q.

a

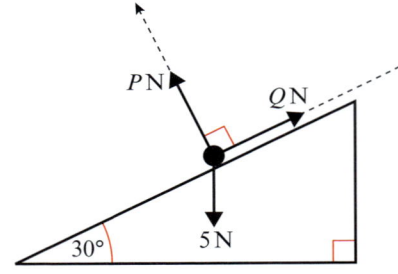

b

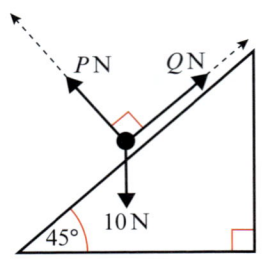

c

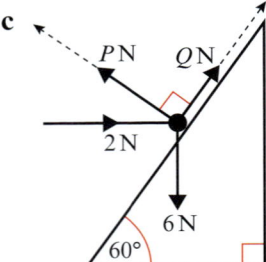

d

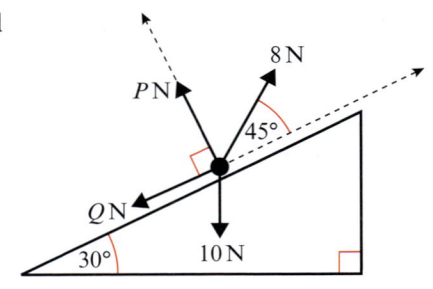

e

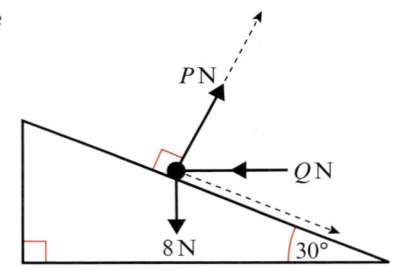

Challenge

The diagram shows three **coplanar** forces of A, B and C acting on a particle in equilibrium.

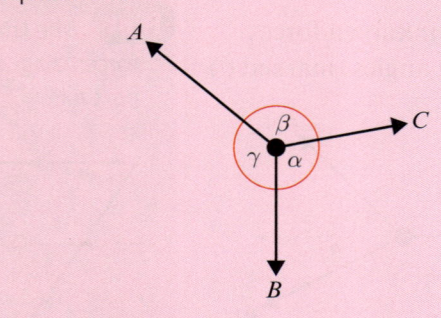

Show that $\dfrac{A}{\sin \alpha} = \dfrac{B}{\sin \beta} = \dfrac{C}{\sin \gamma}$.

Notation This result is known as **Lami's Theorem**.

STATICS OF A PARTICLE CHAPTER 7

7.2 Modelling with statics

You can use force diagrams to model objects in static equilibrium, and to solve problems involving weight, tension and pulleys.

Example 3 — SKILLS: PROBLEM-SOLVING

A smooth bead Y is threaded on a light inextensible string. The ends of the string are attached to two fixed points, X and Z, on the same horizontal level. The bead is held in equilibrium by a horizontal force of magnitude 8 N acting parallel to ZX. The bead Y is vertically below X and $\angle XZY = 30°$ as shown in the diagram.

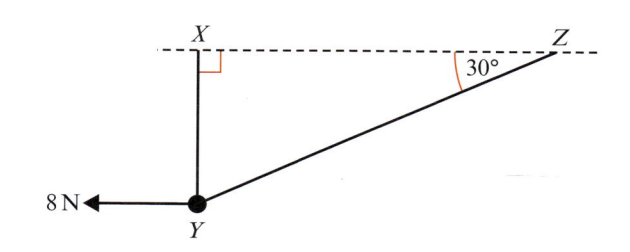

Find the tension in the string and the weight of the bead.

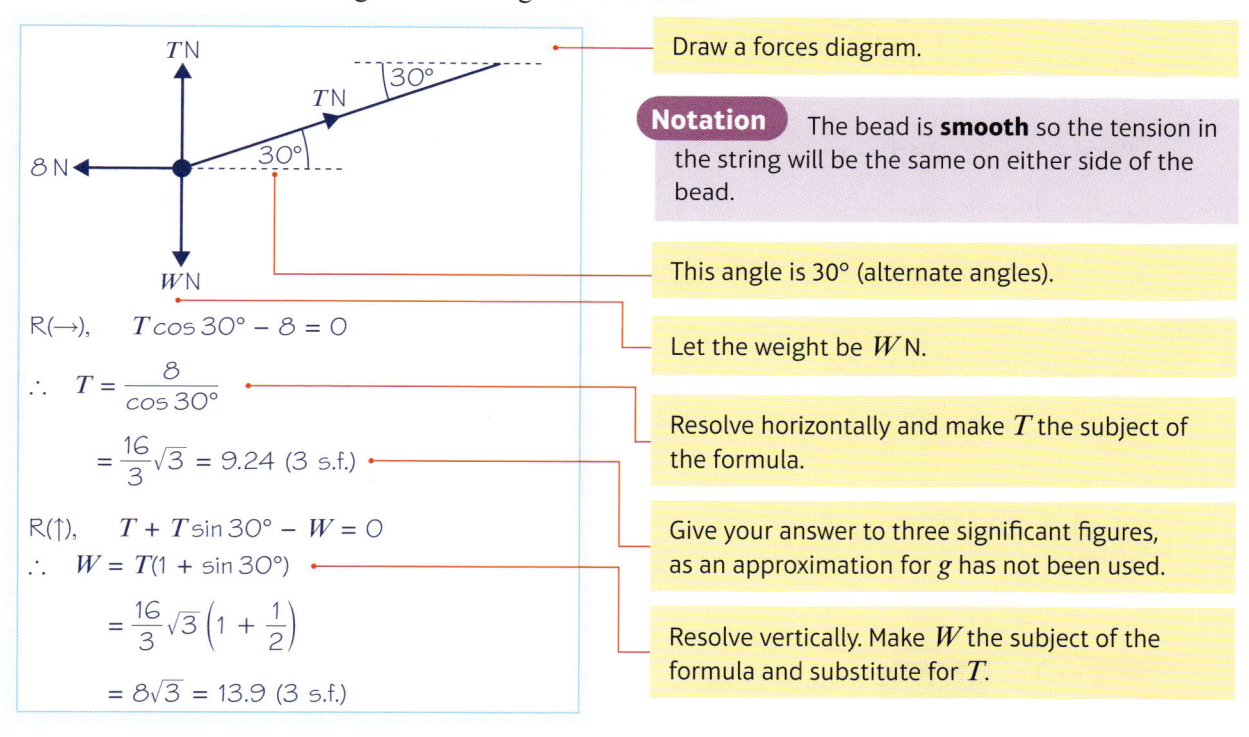

Draw a forces diagram.

Notation The bead is **smooth** so the tension in the string will be the same on either side of the bead.

This angle is 30° (alternate angles).

$R(\rightarrow), \quad T\cos 30° - 8 = 0$

$\therefore T = \dfrac{8}{\cos 30°}$

$= \dfrac{16}{3}\sqrt{3} = 9.24$ (3 s.f.)

Let the weight be W N.

Resolve horizontally and make T the subject of the formula.

$R(\uparrow), \quad T + T\sin 30° - W = 0$

$\therefore W = T(1 + \sin 30°)$

$= \dfrac{16}{3}\sqrt{3}\left(1 + \dfrac{1}{2}\right)$

$= 8\sqrt{3} = 13.9$ (3 s.f.)

Give your answer to three significant figures, as an approximation for g has not been used.

Resolve vertically. Make W the subject of the formula and substitute for T.

Example 4 — SKILLS: PROBLEM-SOLVING

A mass of 3 kg rests on the surface of a smooth plane which is inclined at an angle of 45° to the horizontal. The mass is attached to a cable which passes up the plane along the line of greatest slope and then passes over a smooth pulley at the top of the plane. The cable carries a mass of 1 kg freely suspended at the other end. The masses are modelled as particles, and the cable as a light inextensible string. There is a force of P N acting horizontally on the 3 kg mass and the system is in equilibrium.

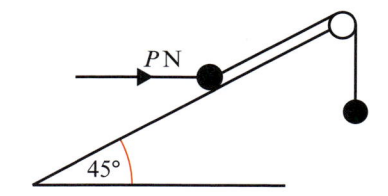

Calculate: **a** the magnitude of P **b** the normal reaction between the mass and the plane.
c State how you have used the assumption that the pulley is smooth in your calculations.

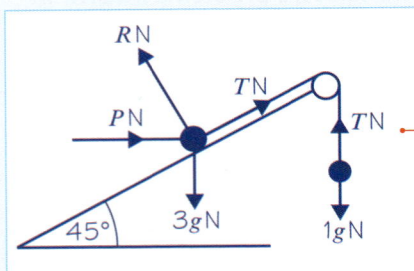

Draw a diagram showing the forces acting on each particle. The tension, T N, will be the same throughout the string. The normal reaction, R N acts perpendicular to the plane. Show the weights $3g$ N and $1g$ N.

a Consider the 1 kg mass:

$R(\uparrow)$, $T - 1g = 0$

Resolve vertically to obtain T.

$\therefore T = g = 9.8$

Consider the 3 kg mass:

$R(\nearrow)$, $T + P\cos 45° - 3g \sin 45° = 0$

Resolve up the plane.

R has no component in this direction as R is perpendicular to the plane.

$\therefore P \cos 45° = 3g \sin 45° - T$

But $T = g$

Substitute the value for T you found earlier.

$\therefore P \cos 45° = 3g \sin 45° - g$

Divide this equation by $\cos 45°$ and use the fact that $\dfrac{\sin 45°}{\cos 45°} = \tan 45° = 1$.

$P = 3g - \dfrac{g}{\cos 45°}$

$= 3g - g\sqrt{2} = 16$ (2 s.f.)

Use the result that $\cos 45° = \sin 45° = \dfrac{1}{\sqrt{2}}$.

b $R(\nwarrow)$, $R - P \sin 45° - 3g \cos 45° = 0$

$\therefore R = P \sin 45° + 3g \cos 45°$

Resolve perpendicular to the plane.

$= 6g \dfrac{\sqrt{2}}{2} - g = 32$ (2 s.f.)

Substitute the value of P which you have found to evaluate R.

c The pulley is smooth so the tension in the string will be the same on both sides of the pulley.

Exercise 7B SKILLS PROBLEM-SOLVING

P **1** A picture of mass 5 kg is suspended by two light inextensible strings, each inclined at 45° to the horizontal as shown. By modelling the picture as a particle, find the tension in the strings when the system is in equilibrium.

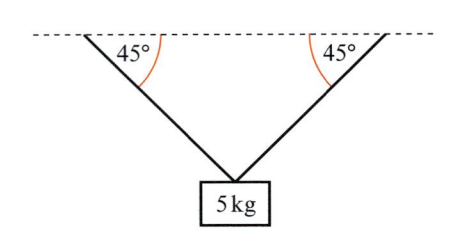

Problem-solving

This is a three-force problem involving an object in static equilibrium, so you could use a triangle of forces.

2 A particle of mass m kg is suspended by a single light inextensible string. The string is inclined at an angle of 30° to the vertical and the other end of the string is attached to a fixed point O. Equilibrium is maintained by a horizontal force of magnitude 10 N which acts on the particle, as shown in the diagram. Find:
 a the tension in the string **b** the value of m.

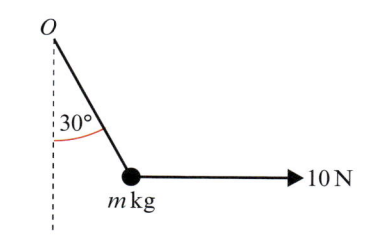

3 A particle of weight 12 N is suspended by a light inextensible string from a fixed point O. A horizontal force of 8 N is applied to the particle, and the particle remains in equilibrium with the string at an angle θ to the vertical. Find:
 a the angle θ **b** the tension in the string.

4 A particle of mass 6 kg hangs in equilibrium, suspended by two light inextensible strings, inclined at 60° and 45° to the horizontal, as shown. Find the tension in each of the strings.

Hint The particle is attached **separately** to each string, so the tension in the two strings can be different.

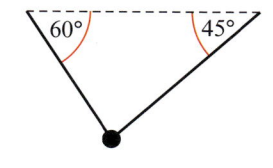

(E) 5 A smooth bead B is threaded on a light inextensible string. The ends of the string are attached to two fixed points, A and C, on the same horizontal level. The bead is held in equilibrium by a horizontal force of magnitude 2 N acting parallel to CA. The sections of string make angles of 60° and 30° with the horizontal. Find:

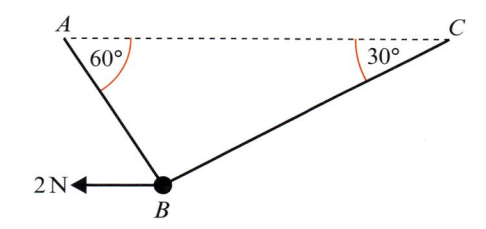

 a the tension in the string (3 marks)

 b the mass of the bead. (4 marks)

 c State how you have used the modelling assumption that the bead is smooth in your calculations. (1 mark)

(E) 6 A smooth bead B is threaded on a light inextensible string. The ends of the string are attached to two fixed points A and C where A is vertically above C. The bead is held in equilibrium by a horizontal force of magnitude 2 N. The sections AB and BC of the string make angles of 30° and 60° with the vertical respectively. Find:

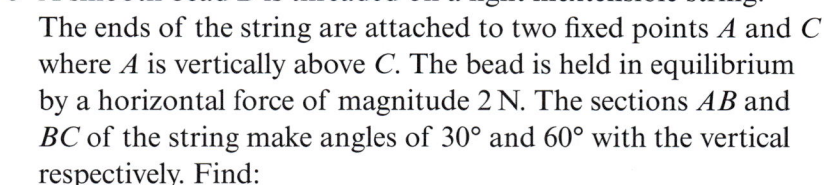

 a the tension in the string (3 marks)

 b the mass of the bead, giving your answer to the nearest gram. (4 marks)

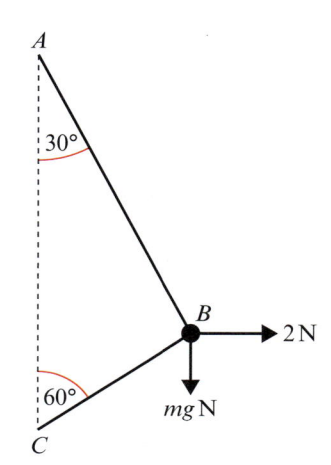

(E) 7 A particle of weight 2 N rests on a smooth horizontal surface and remains in equilibrium under the action of the two external forces shown in the diagram. One is a horizontal force of magnitude 1 N and the other is a force of magnitude P N which acts at an angle θ to the horizontal, where $\tan\theta = \frac{12}{5}$. Find:

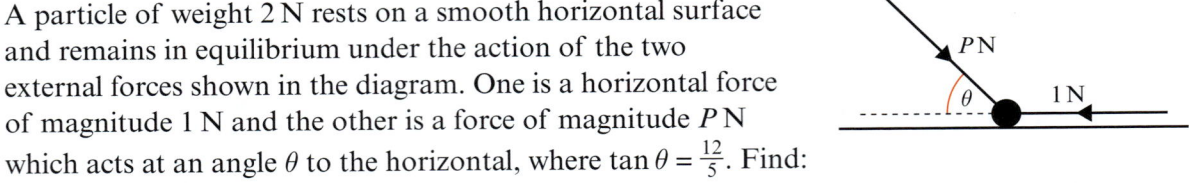

a the value of P (3 marks)

b the normal reaction between the particle and the surface. (2 marks)

(P) 8 A particle A of mass m kg rests on a smooth horizontal table. The particle is attached by a light inextensible string to another particle B of mass $2m$ kg, which hangs over the edge of the table. The string passes over a smooth pulley, which is fixed at the edge of the table so that the string is horizontal between A and the pulley and then is vertical between the pulley and B. A horizontal force F N applied to A maintains equilibrium. The normal reaction between A and the table is R N.

a Find the values of F and R in terms of m.

The pulley is now raised to a position above the edge of the table so that the string is inclined at 30° to the horizontal between A and the pulley. The string still hangs vertically between the pulley and B. A horizontal force F' N applied to A maintains equilibrium in this new situation. The normal reaction between A and the table is now R' N.

b Find, in terms of m, the values of F' and R'.

9 A particle of mass 2 kg rests on a smooth inclined plane, which makes an angle of 45° with the horizontal. The particle is maintained in equilibrium by a force P N acting up the line of greatest slope of the inclined plane, as shown in the diagram. Find the value of P.

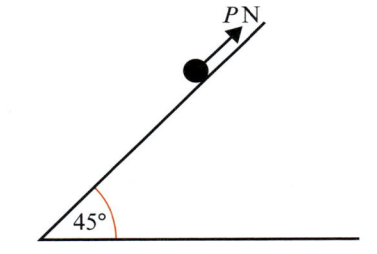

10 A particle of mass 4 kg is held in equilibrium on a smooth plane which is inclined at 45° to the horizontal by a horizontal force of magnitude P N, as shown in the diagram. Find the value of P.

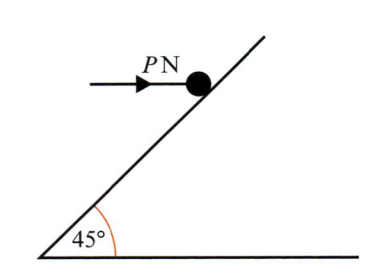

(E/P) 11 A particle A of mass 2 kg rests in equilibrium on a smooth inclined plane. The plane makes an angle θ with the horizontal, where $\tan\theta = \frac{3}{4}$.

The particle is attached to one end of a light inextensible string which passes over a smooth pulley, as shown in the diagram. The other end of the string is attached to a particle B of mass 5 kg. Particle A is also acted upon by a force of magnitude F N down the plane, along the line of greatest slope.

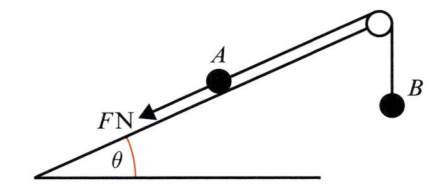

Find:
  a the magnitude of the normal reaction between A and the plane (5 marks)
  b the value of F. (3 marks)
  c State how you have used the fact that the pulley is smooth in your calculations. (1 mark)

(E/P) 12 A particle of weight 20 N rests in equilibrium on a smooth inclined plane. It is maintained in equilibrium by the application of two external forces as shown in the diagram. One of the forces is a horizontal force of 5 N, the other is a force P N acting at an angle of 30° to the plane, as shown in the diagram. Find the magnitude of the normal reaction between the particle and the plane. (8 marks)

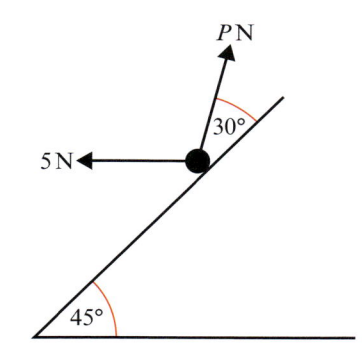

7.3 Friction and static particles

When a body is in static equilibrium under the action of a number of forces, including friction, you need to consider whether or not the body is on the point of moving.

In many cases the force of friction will be less than μR, as a smaller force is sufficient to prevent motion and to maintain static equilibrium. In these situations the equilibrium is not limiting.

- **The maximum value of the frictional force $F_{MAX} = \mu R$ is reached when the body you are considering is on the point of moving. The body is then said to be in limiting equilibrium.**

- **In general, the force of friction F is such that $F \leq \mu R$, and the direction of the frictional force is opposite to the direction in which the body would move if the frictional force were absent.**

Example 5 SKILLS PROBLEM-SOLVING

A mass of 8 kg rests on a rough horizontal plane. The mass may be modelled as a particle, and the coefficient of friction between the mass and the plane is 0.5. Find the magnitude of the maximum force P N which acts on this mass without causing it to move if:
a the force P is horizontal
b the force P acts at an angle 60° above the horizontal.

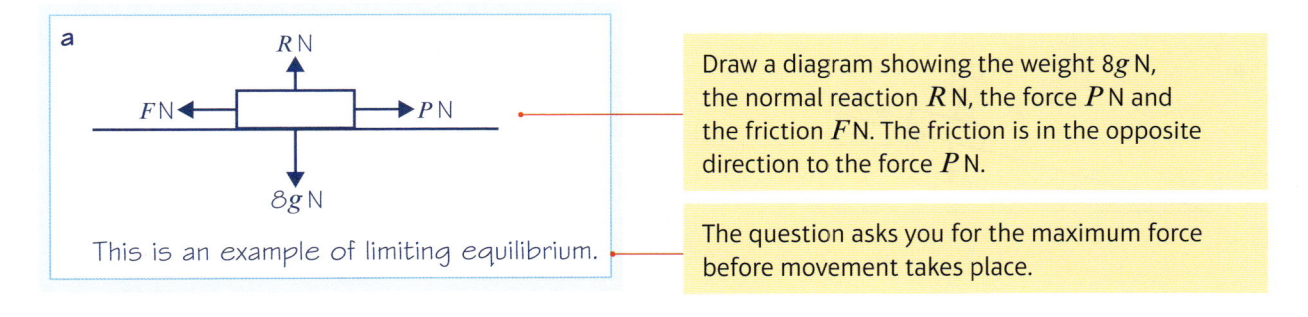

a

This is an example of limiting equilibrium.

Draw a diagram showing the weight $8g$ N, the normal reaction R N, the force P N and the friction F N. The friction is in the opposite direction to the force P N.

The question asks you for the maximum force before movement takes place.

$R(\uparrow)$, $\quad R - 8g = 0$
$\therefore \quad R = 8g$
As friction is limiting, $F = \mu R$ — *For an object in limiting equilibrium, $F = F_{MAX}$.*
$\therefore \quad F = 0.5 \times 8g$
$\quad \quad = 39.2\,N$
$R(\rightarrow)$, $\quad P - F = 0$
$\therefore \quad P = F = 39\,N$ (2 s.f.) — *Give your answer to two significant figures.*

b — *Draw another diagram showing P at 60° above the horizontal.*

Again this is limiting equilibrium.
$R(\uparrow)$, $\quad R + P\sin 60° - 8g = 0$
$\therefore \quad R = 8g - P\sin 60°$ — *Express R in terms of P.*
As friction is limiting, $F = \mu R$
$\therefore \quad F = 0.5\,(8g - P\sin 60°)$ — *Use $F = \mu R$ with $\mu = 0.5$.*
$R(\rightarrow)$, $\quad P\cos 60° - F = 0$
$\therefore \quad P\cos 60° = 0.5\,(8g - P\sin 60°)$ — *As $F = P\cos 60°$ eliminate F from the previous equation.*
$\therefore \quad P\cos 60° + 0.5\,P\sin 60° = 0.5 \times 8g$
$\therefore \quad P(\cos 60° + 0.5\sin 60°) = 4g$ — *Collect the terms in P and factorise to make P the subject.*
$\therefore \quad P = \dfrac{4g}{\cos 60° + 0.5\sin 60°}$
$P = 42.0\,N$ (3 s.f.)

Example 6 — SKILLS PROBLEM-SOLVING

A box of mass 10 kg rests in limiting equilibrium on a rough plane inclined at 20° above the horizontal.

a Find the coefficient of friction between the box and the plane.

A horizontal force of magnitude P N is applied to the box.

b Given that the box remains in equilibrium, find the maximum possible value of P.

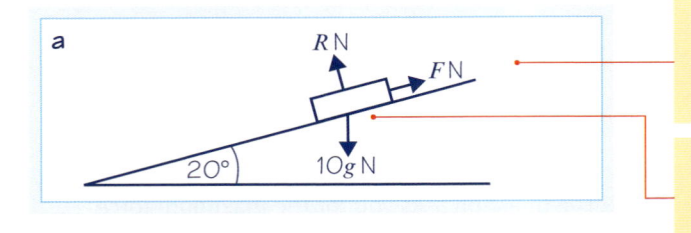

a — *Model the box as a particle and draw a diagram showing the weight, the normal reaction and the force of friction.*

The friction acts up the plane, as it acts in an opposite direction to the motion that would take place if there was no friction.

STATICS OF A PARTICLE — CHAPTER 7

$R(\nwarrow),\quad R - 10g\cos 20° = 0$
$\therefore R = 92.089\ldots \text{N}$

$R(\nearrow),\quad F - 10g\sin 20° = 0$
$\therefore F = 33.517\ldots \text{N}$

Resolve perpendicular and parallel to the plane.

Online Use the **STO** function to store exact values on your calculator.

As the friction is limiting,
$F = \mu R$
$\therefore 33.517\ldots = \mu \times 92.089\ldots$
$\therefore \mu = \dfrac{33.517\ldots}{92.089\ldots} = 0.36 \text{ (to 2 s.f.)}$

Find R and F, then use $F = \mu R$ to find μ.

Give your answer to two significant figures and note that $\mu = \tan 20°$.

b

Diagram: block on slope at 20°, forces P N (horizontal into slope), R N (normal), $0.36R$ (friction down slope), $10g$ N (weight).

Watch out For the maximum possible value of P, the box will be on the point of moving up the slope, so the friction will act down the slope.

When P is at its maximum value,
$F = \mu R = 0.36R$

$R(\nearrow),\ P\cos 20° - 0.36R - 10g\sin 20° = 0$ (1)

$R(\nwarrow),\ R - 10g\cos 20° - P\sin 20° = 0$ (2)

From (1): $R = \dfrac{P\cos 20° - 10g\sin 20°}{0.36}$

From (2) $R = P\sin 20° + 10g\cos 20°$

So $\dfrac{P\cos 20° - 10g\sin 20°}{0.36} = P\sin 20° + 10g\cos 20°$

Eliminate R to find P.

$P = \dfrac{3.6g\cos 20° + 10g\sin 20°}{\cos 20° - 0.36\sin 20°}$

$= 82 \text{ N (2 s.f.)}$

You have used $g = 9.8 \text{ m s}^{-2}$ in your calculations, so round your final answer to 2 significant figures.

Exercise 7C — SKILLS: PROBLEM-SOLVING

1 A book of mass 2 kg rests on a rough horizontal table. When a force of magnitude 8 N acts on the book, at an angle of 20° to the horizontal in an upward direction, the book is on the point of slipping.

Calculate, to three significant figures, the value of the coefficient of friction between the book and the table.

Hint 'On the point of slipping' means that the book is in limiting equilibrium.

2 A block of mass 4 kg rests on a rough horizontal table. When a force of 6 N acts on the block, at an angle of 30° to the horizontal in a downward direction, the block is on the point of slipping. Find the value of the coefficient of friction between the block and the table.

3 A block of weight 10 N is at rest on a rough horizontal surface. A force of magnitude 3 N is applied to the block at an angle of 60° above the horizontal in an upward direction. The coefficient of friction between the block and the surface is 0.3.
  a Calculate the force of friction.
  b Determine whether or not the friction is limiting.

4 A packing crate of mass 10 kg rests on rough horizontal ground. It is filled with books which are evenly distributed through the crate. The coefficient of friction between the crate and the ground is 0.3.
  a Find the mass of the books if the crate is in limiting equilibrium under the effect of a horizontal force of magnitude 147 N.
  b State what modelling assumptions you have made.

5 A block of mass 2 kg rests on a rough horizontal plane. A force P acts on the block at an angle of 45° to the horizontal. The equilibrium is limiting, with $\mu = 0.3$.

  Find the magnitude of P if:
  a P acts in a downward direction
  b P acts in an upward direction.

6 A particle of mass 0.3 kg is on a rough plane which is inclined at an angle of 30° to the horizontal. The particle is held at rest on the plane by a force of magnitude 3 N acting up the plane, in a direction parallel to the line of greatest slope of the plane. The particle is on the point of slipping up the plane. Find the coefficient of friction between the particle and the plane.

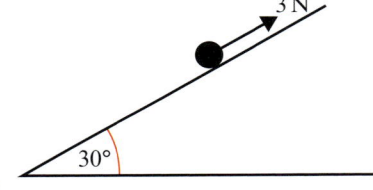

7 A particle of mass 1.5 kg rests in equilibrium on a rough plane under the action of a force of magnitude X N acting up the line of greatest slope of the plane. The plane is inclined at 25° to the horizontal. The particle is in limiting equilibrium and on the point of moving up the plane. The coefficient of friction between the particle and the plane is 0.25. Calculate:
  a the normal reaction of the plane on the particle 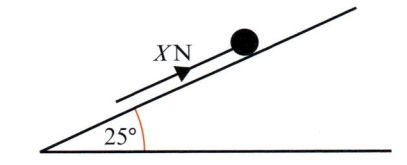
  b the value of X.

8 A horizontal force of magnitude 20 N acts on a block of mass 1.5 kg, which is in equilibrium resting on a rough plane inclined at 30° to the horizontal. The line of action of the force is in the same vertical plane as the line of greatest slope of the inclined plane. 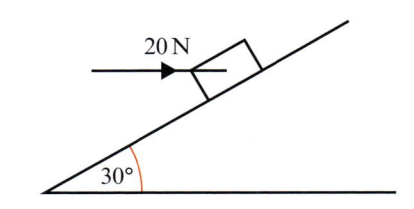
  a Find the normal reaction between the block and the plane. (4 marks)
  b Find the magnitude and direction of the frictional force acting on the block. (3 marks)
  c Hence find the **minimum** value of the coefficient of friction between the block and the plane. (2 marks)

9 A box of mass 3 kg lies on a rough plane inclined at 40° to the horizontal. The box is held in equilibrium by means of a horizontal force of magnitude X N. The line of action of the force is in the same vertical plane as the line of greatest slope of the inclined plane. The coefficient of friction between the box and the plane is 0.3 and the box is in limiting equilibrium and is about to move up the plane.

  a Find X. **(6 marks)**

  b Find the normal reaction between the box and the plane. **(2 marks)**

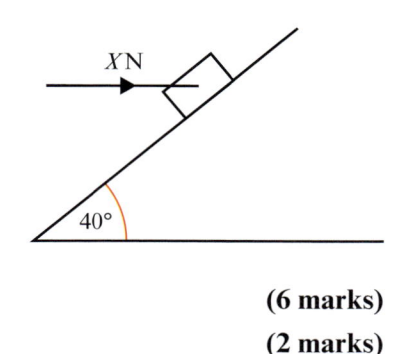

10 A small child, sitting on a sledge, rests in equilibrium on an inclined slope. The sledge is held by a rope which lies along the slope and is under tension. The sledge is on the point of slipping down the plane. Modelling the child and sledge as a particle and the rope as a light inextensible string, calculate the tension in the rope, given that the combined mass of the child and sledge is 22 kg, the coefficient of friction is 0.125, and that the slope is a plane inclined at 35° to the horizontal.

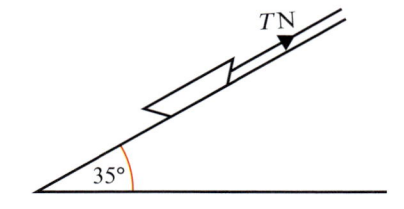

11 A box of mass 0.5 kg is placed on a plane which is inclined at an angle of 40° to the horizontal. The coefficient of friction between the box and the plane is $\frac{1}{5}$. The box is kept in equilibrium by a light string which lies in a vertical plane containing a line of greatest slope of the plane. The string makes an angle of 20° with the plane, as shown in the diagram. The box is in limiting equilibrium and may be modelled as a particle. The tension in the string is T N.

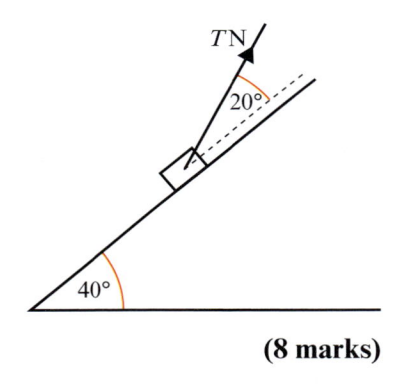

  Find the range of possible values of T. **(8 marks)**

  Problem-solving
  The box might be about to move up or down the slope.

12 A box of mass 1 kg is placed on a plane, which is inclined at an angle of 40° to the horizontal. The box is kept in equilibrium on the point of moving up the plane by a light string, which lies in a vertical plane containing a line of greatest slope of the plane. The string makes an angle of 20° with the plane, as shown in the diagram. The box is in limiting equilibrium and may be modelled as a particle. The tension in the string is 10 N and the coefficient of friction between the box and the plane is μ. Find μ. **(7 marks)**

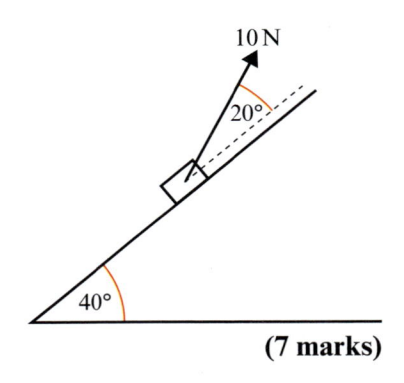

E/P 13 A box of mass 2 kg rests in limiting equilibrium on a rough plane angled at θ above the horizontal where $\tan\theta = \frac{3}{4}$. A horizontal force of magnitude P N acting into the plane is applied to the box. Given that the box remains in equilibrium, find the maximum possible value of P. **(8 marks)**

Problem-solving
First find the coefficient of friction between the box and the plane.

Chapter review 7

1 A particle is acted upon by three forces as shown in the diagram.

Given that the particle is in equilibrium, work out:
  a the size of angle θ
  b the magnitude of P.

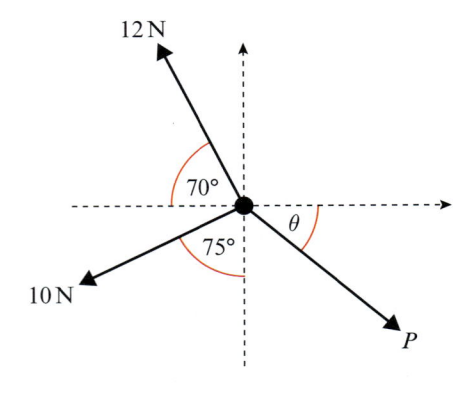

2 A particle is acted upon by three forces as shown in the diagram. Given that it is in equilibrium find:
  a the size of angle θ
  b the magnitude of W.

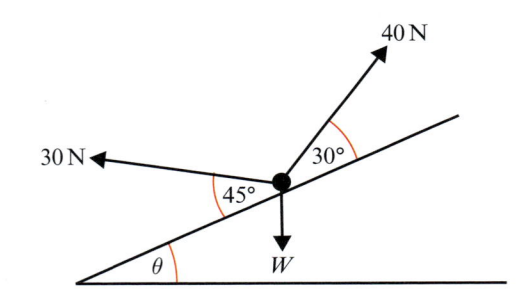

P 3 An acrobat of mass 55 kg stands on a tightrope. By modelling the acrobat as a particle and the tightrope as two inextensible strings, calculate the tension in the tightrope on each side of the rope.

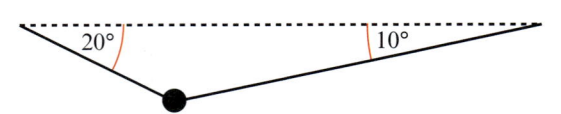

E 4 A particle Q of mass 5 kg rests in equilibrium on a smooth inclined plane. The plane makes an angle of θ with the horizontal, where $\tan\theta = \frac{3}{4}$.
Q is attached to one end of a light inextensible string which passes over a smooth pulley as shown. The other end of the string is attached to a particle of mass 2 kg.

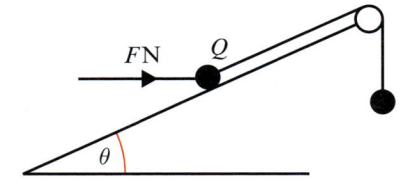

The particle Q is also acted upon by a force of magnitude F N acting horizontally, as shown in the diagram.

Find the magnitude of:

a the force F **(5 marks)**

b the normal reaction between particle Q and the plane. **(3 marks)**

The plane is now assumed to be rough.

c State, with a reason, which of the following statements is true:
  1. F will be larger 2. F will be smaller 3. F could be either larger or smaller. **(2 marks)**

(E) **5** A smooth bead B of mass 2 kg is threaded on a light inextensible string. The ends of the string are attached to two fixed points A and C, where A is vertically above C. The bead is held in equilibrium by a horizontal force of magnitude P N. The sections AB and BC make angles of 20° and 70° with the vertical as shown.

a Show that the tension in the string is 33 N (2 s.f.). **(3 marks)**

b Calculate the value of P. **(3 marks)**

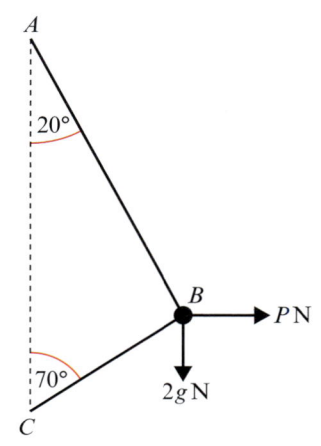

(P) **6** A box of mass m kg sits stationary on a rough plane. The plane is inclined at an angle of θ to the horizontal and has coefficient of friction μ. Show that when the box is on the point of slipping $\theta = \tan^{-1}\mu$.

(P) **7** A particle of mass m kg lies at rest on a rough horizontal surface. A force of magnitude $\frac{1}{5}mg$ is applied to the particle at an angle of θ to the horizontal. Given that the particle remains at rest, show that $\mu \geq \dfrac{\cos\theta}{5 + \sin\theta}$.

8 A particle of mass 5 kg is stationary on a rough horizontal plane. A force of 5 N is applied at an angle of $\tan^{-1}\left(\frac{4}{3}\right)$ to the plane. Given that the particle is on the point of slipping, show that $\mu = \frac{1}{15}$.

9 A box of mass M kg sits stationary on a smooth plane. The plane is inclined at an angle of θ to the horizontal where $\theta = \tan^{-1}\left(\frac{5}{12}\right)$. The box of mass M kg is attached to a second box of mass m kg via an inextensible rope that hangs over a smooth pulley as shown. Show that if the system is to remain at rest, then $\dfrac{m}{M} = \dfrac{5}{13}$.

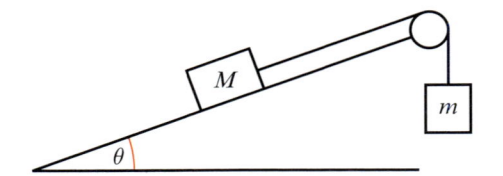

Summary of key points

1. A particle or rigid body is in static equilibrium if it is at rest and the resultant force acting on the particle or rigid body is zero.

2. The maximum value of the frictional force $F_{MAX} = \mu R$ is reached when the body you are considering is on the point of moving. The body is then said to be in limiting equilibrium.

3. In general, the force of friction F is such that $F \leq \mu R$, and the direction of the frictional force is opposite to the direction in which the body would move if the frictional force were absent.

4. For a rigid body in static equilibrium:
   - the body is stationary
   - the resultant force in any direction is zero
   - the resultant moment is zero.

8 MOMENTS

6.1

Learning objectives

After completing this chapter you should be able to:
- Calculate the turning effect of a force applied to a rigid body → pages 130–131
- Calculate the resultant moment of a set of forces acting on a rigid body → pages 132–133
- Solve problems involving uniform rods in equilibrium → pages 133–136
- Solve problems involving non-uniform rods → pages 136–139
- Solve problems involving rods on the point of tilting → pages 139–141

Prior knowledge check

1 Find the value of x in each of the following:

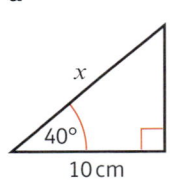

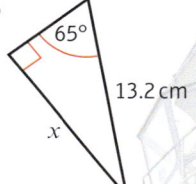

   a right triangle with angle 40°, base 10 cm, hypotenuse x

   b right triangle with angle 65°, hypotenuse 13.2 cm, side x

   ← International GCSE Mathematics

2 Masses A and B rest on a light scale-pan supported by two strings, each with tension T.

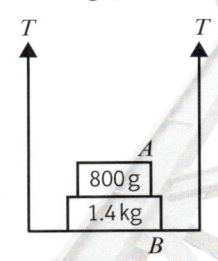

   (Diagram: two upward tensions T supporting a pan with mass A = 800 g on top and mass B = 1.4 kg below)

   Find:
   a the value of T
   b the normal reaction of the scale-pan on mass B
   c the normal reaction of mass B on mass A.

   ← Mechanics 1 Section 4.5

Moments measure the turning effect of a force. Engineers use moments to work out how much load can be applied safely to a crane.

8.1 Moments

So far you have looked mostly at situations involving particles. This means you can ignore rotational effects. In this chapter, you begin to model objects as **rigid bodies**. This allows you to consider the size of the object as well as where forces are applied.

The **moment** of a force measures the turning effect of the force on a rigid body.

It is the product of the magnitude of the force and the perpendicular distance from the axis of rotation.

- Clockwise moment of F about $P = |F| \times d$

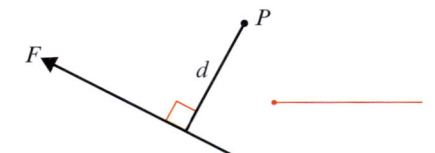

The moment of the force, F, is acting about the point P.

Watch out When you describe a moment, you need to give the direction of rotation.

In the diagram above, the distance given is perpendicular to the line of action of the force. When this is not the case, you need to use trigonometry to find the perpendicular distance.

- Clockwise moment of F about $P = |F| \times d \sin \theta$

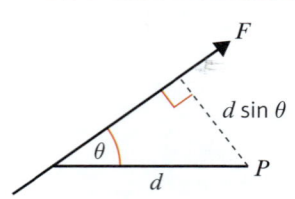

Notation A moment is a force multiplied by a distance, so its units are **newton metres (N m)** or **newton centimetres (N cm)**.

Online Explore the moment of a force acting about a point using GeoGebra.

Example 1   SKILLS   PROBLEM-SOLVING

Find the moment of each force about the point P.

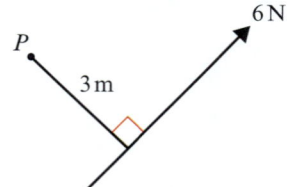

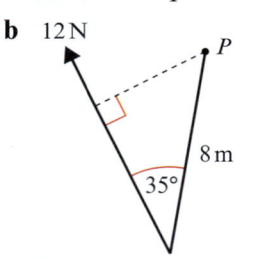

a 6 N, 3 m
b 12 N, 8 m, 35°

a Moment of the 6 N force about P
= magnitude of force × perpendicular distance
= 6 × 3 = 18 N m anticlockwise

b Moment of the 12 N force about P
= magnitude of force × perpendicular distance
= 12 × 8 sin 35° = 55.1 N m clockwise (3 s.f.)

The distance given on the diagram is the perpendicular distance, so you can substitute the given values directly into the formula.

Don't forget to include the direction of the rotation when you describe the moment of the force.

This time you need to use the perpendicular distance 8 sin 35°.

MOMENTS   CHAPTER 8   131

Example 2   SKILLS PROBLEM-SOLVING

The diagram shows two forces acting on a lamina. Find the moment of each of the forces about the point P.

Moment of the 5 N force
= magnitude of force × perpendicular distance
= 5 × 2 = 10 N m clockwise

Moment of the 8 N force
= magnitude of force × perpendicular distance
= 8 × 2 sin 50° = 12.3 N m anticlockwise (3 s.f.)

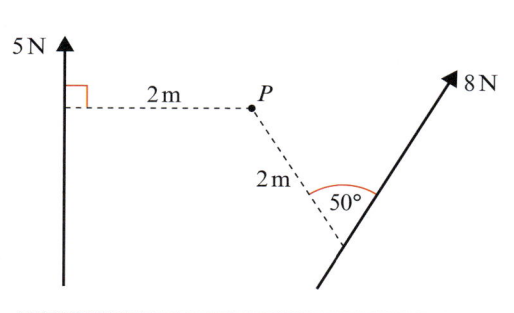

Notation A lamina is a 2D object whose thickness can be ignored.

The moments act in opposite directions.

Exercise 8A   SKILLS PROBLEM-SOLVING

1 Calculate the moment about P of each of these forces acting on a lamina.

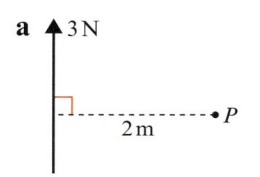

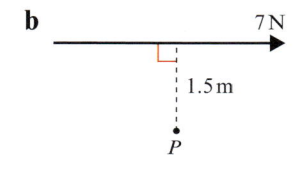

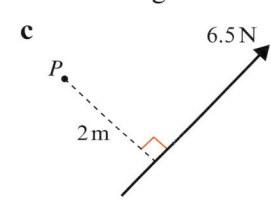

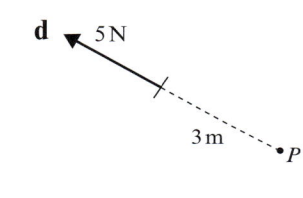

  a 3 N, 2 m   **b** 7 N, 1.5 m   **c** 6.5 N, 2 m   **d** 5 N, 3 m

2 Calculate the moment about P of each of these forces acting on a lamina.

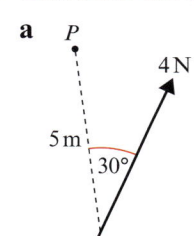

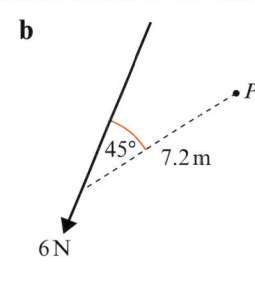

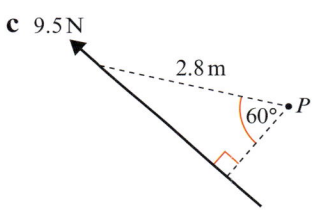

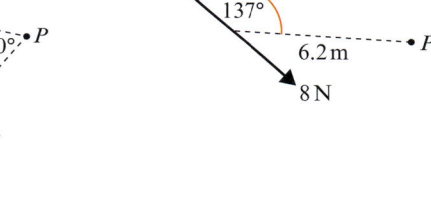

  a 4 N, 5 m, 30°   **b** 6 N, 7.2 m, 45°   **c** 9.5 N, 2.8 m, 60°   **d** 8 N, 6.2 m, 137°

3 The diagram shows a sign hanging from a wooden beam. The sign has a mass of 4 kg.

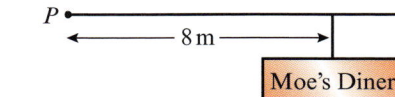

  a Calculate the moment of the weight of the mass:

   i about P **ii** about Q.

  b Comment on any modelling assumptions you have made.

(P) **4** $ABCD$ is a rectangular lamina. A force of 12 N acts horizontally at B, as shown in the diagram. Find the moment of this force about:

  a A **b** B **c** C **d** D

(P) **5** In the diagram, the force **F** produces a moment of 15 N m clockwise about the **pivot** P. Calculate the magnitude of **F**.

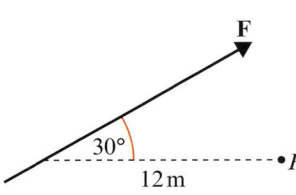

8.2 Resultant moments

When you have several **coplanar** forces acting on a body, you can determine the turning effect around a given point by choosing a positive direction (clockwise or anticlockwise) and then finding the sum of the moments produced by each force.

Notation Coplanar forces are forces that act in the same plane.

- The sum of the moments acting on a body is called the resultant moment.

Example 3 SKILLS PROBLEM-SOLVING

The diagram shows a set of forces acting on a light rod. Calculate the resultant moment about the point P.

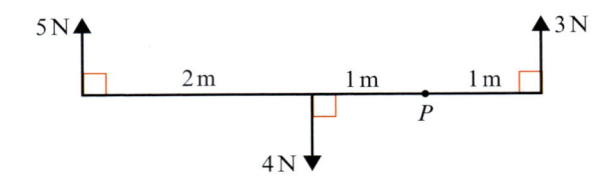

The moment of the 5 N force is
$5 \times (2 + 1) = 15$ N m clockwise.

The moment of the 4 N force is
$4 \times 1 = 4$ N m anticlockwise.

The moment of the 3 N force is
$3 \times 1 = 3$ N m anticlockwise.

Choosing clockwise as positive:

Resultant moment = $15 + (-4) + (-3) = 8$ N m

∴ resultant moment is 8 N m clockwise.

Your positive direction is clockwise, so the anticlockwise moments are negative.

Problem-solving

You could also solve this problem by considering the clockwise and anticlockwise moments separately.
Sum of clockwise moments = 15 N m
Sum of anticlockwise moments = $3 + 4 = 7$ N m
Resultant moment = $15 - 7 = 8$ N m clockwise

Exercise 8B SKILLS PROBLEM-SOLVING

1 These diagrams show sets of forces acting on a light rod. In each case, calculate the resultant moment about P.

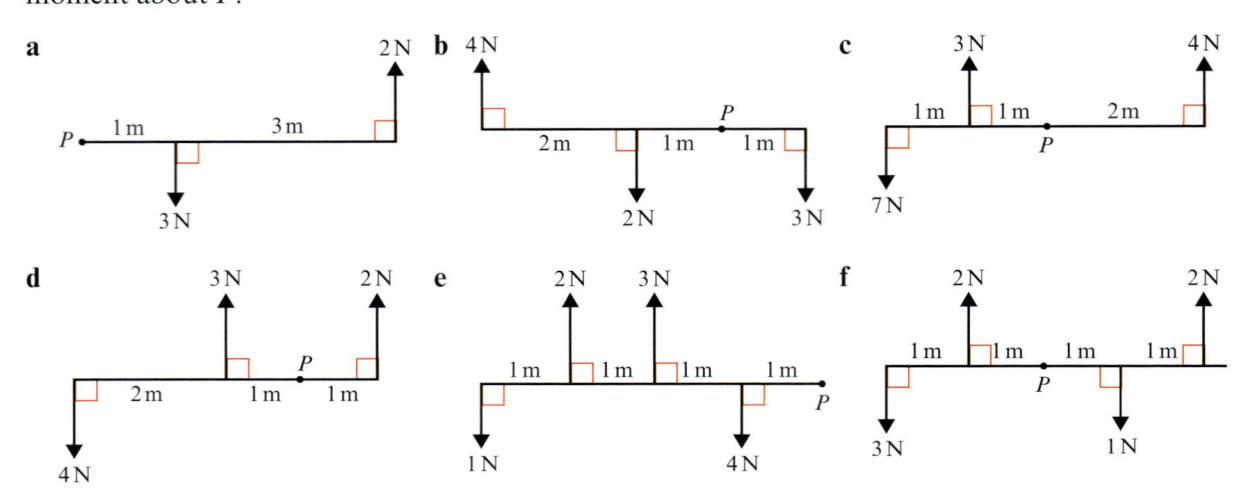

2 These diagrams show forces acting on a lamina. In each case, find the resultant moment about P.

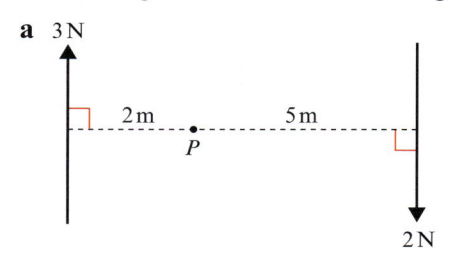

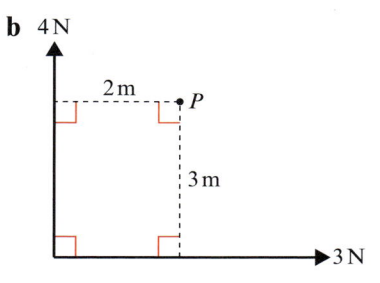

a 3 N, 2 m, P, 5 m, 2 N

b 4 N, 2 m, P, 3 m, 3 N

3 The diagram shows a set of forces acting on a light rod. The resultant moment about P is 17 N m clockwise. Find the length d.

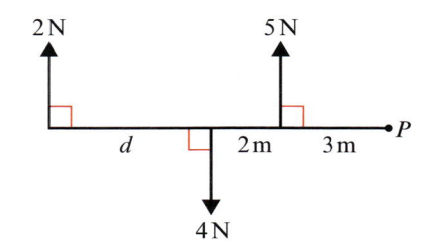

2 N, 5 N, d, 2 m, 3 m, P, 4 N

4 The diagram shows a set of forces acting on a light rod. The resultant moment about P is 12.8 N m clockwise. Find the value of x.

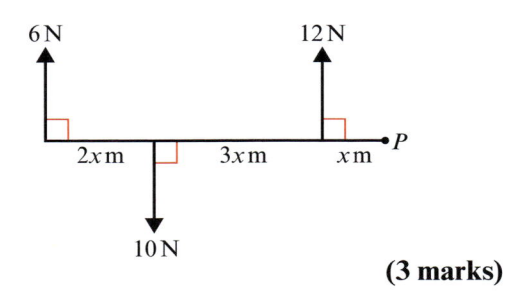

6 N, 12 N, $2x$ m, $3x$ m, x m, P, 10 N

(3 marks)

8.3 Equilibrium

- When a rigid body is in equilibrium, the resultant force in any direction is 0 N and the resultant moment about any point is 0 N m.

Hint If the resultant moment is zero then the sum of the clockwise moments equals the sum of the anticlockwise moments.

You can simplify many problems involving rigid bodies by choosing which point(s) to take moments about. When you take moments at a given point, you can ignore the rotational effect of any forces acting at that point.

Example 4 SKILLS PROBLEM-SOLVING

The diagram shows a uniform rod AB, of length 3 m and weight 20 N, resting horizontally on supports at A and C, where AC = 2 m.

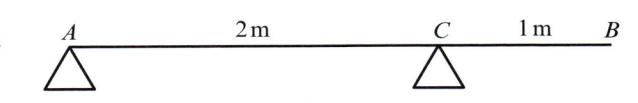

Calculate the magnitude of the reaction at each of the supports.

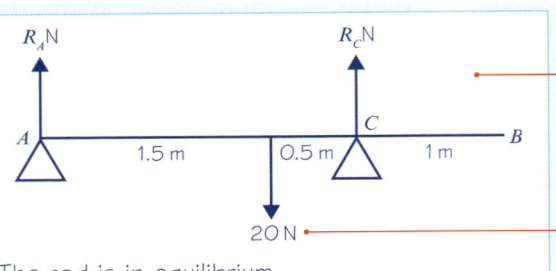

CHAPTER 8 — MOMENTS

The rod is in equilibrium.
Resolving vertically: $R_A + R_C = 20$
Considering the moments about point A:
$20 \times 1.5 = R_C \times (1.5 + 0.5)$
$30 = 2R_C$
$15 = R_C$
$R_A + 15 = 20$
$R_A = 5$
Therefore the reaction at A is 5 N and the reaction at C is 15 N.

- Draw a diagram showing all the forces acting.
- The weight of the rod acts at its centre of mass. You are told that this is a uniform rod, so the weight acts at the midpoint of the rod.
- Total of forces acting upward = total of forces acting downward.
- Clockwise moment = anticlockwise moment

Problem-solving

Take moments about the point that makes the algebra as simple as possible. Taking moments about A results in an equation with just one unknown.

Substituting the value of R_C into the first equation.

Example 5   SKILLS   PROBLEM-SOLVING

A uniform beam AB, of mass 40 kg and length 5 m, rests horizontally on supports at C and D, where $AC = DB = 1$ m. When a man of mass 80 kg stands on the beam at E the magnitude of the reaction at D is twice the magnitude of the reaction at C. By modelling the beam as a rod and the man as a particle, find the distance AE.

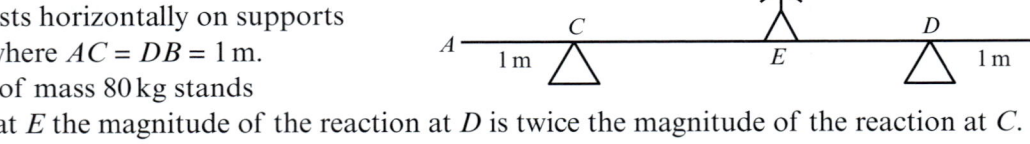

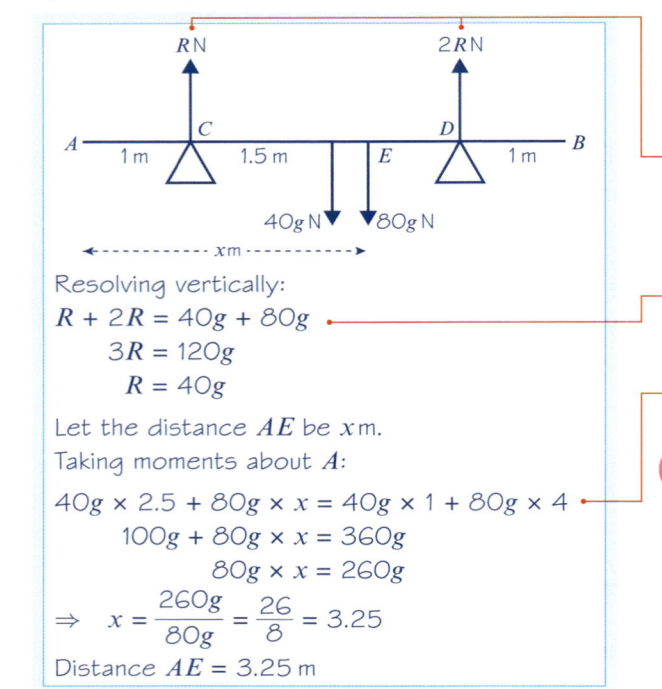

Resolving vertically:
$R + 2R = 40g + 80g$
$3R = 120g$
$R = 40g$

Let the distance AE be x m.
Taking moments about A:
$40g \times 2.5 + 80g \times x = 40g \times 1 + 80g \times 4$
$100g + 80g \times x = 360g$
$80g \times x = 260g$
$\Rightarrow x = \dfrac{260g}{80g} = \dfrac{26}{8} = 3.25$
Distance $AE = 3.25$ m

- Draw a diagram showing the forces.
- Because you are told a relationship between the reaction at C and the reaction at D you can use this on your diagram.
- The rod is in equilibrium so there is no resultant force.
- Clockwise moment = anticlockwise moment

Problem-solving

How have you used the modelling assumptions in the question?
- Since the beam is a rod, it does not bend.
- Since the man is a particle, his weight acts at the point E.

MOMENTS   CHAPTER 8

Exercise 8C   SKILLS PROBLEM-SOLVING

1 AB is a uniform rod of length 5 m and weight 20 N. In these diagrams AB is resting in a horizontal position on supports at C and D. In each case, find the magnitudes of the reactions at C and D.

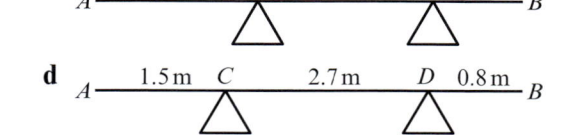

   a A — 1 m — C — 3 m — D — 1 m — B

   b A — 2 m — C — 2 m — D — 1 m — B

   c A — 1.5 m — C — 2.5 m — D — 1 m — B

   d A — 1.5 m — C — 2.7 m — D — 0.8 m — B

2 Each of these diagrams shows a light rod in equilibrium in a horizontal position under the action of a set of forces. Find the values of the unknown forces and distances.

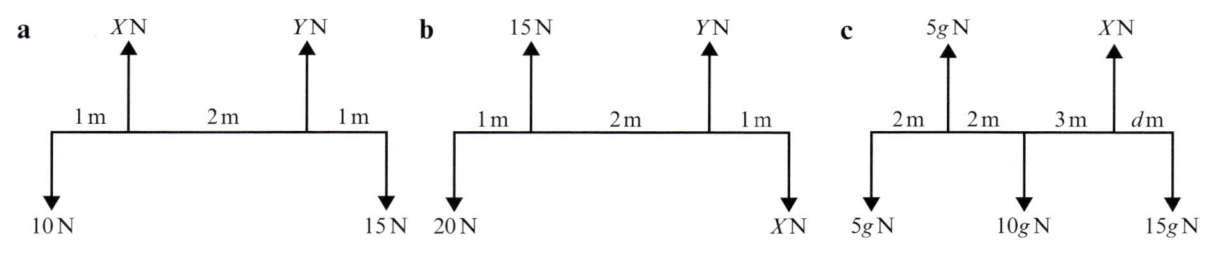

   a Upward forces: X N, Y N; Downward force: 10 N. Distances: 1 m, 2 m, 1 m.

   b Upward forces: 15 N, Y N; Downward forces: 15 N, 20 N. Distances: 1 m, 2 m, 1 m.

   c Upward forces: $5g$ N, X N; Downward forces: X N, $5g$ N, $10g$ N, $15g$ N. Distances: 2 m, 2 m, 3 m, d m.

3 Jack and Jill are playing on a seesaw made from a uniform plank AB, of length 5 m pivoted at M, the midpoint of AB. Jack has mass 35 kg and Jill has mass 28 kg. Jill sits at A and Jack sits at a distance x m from B. The plank is in equilibrium. Find the value of x.

4 A uniform rod AB, of length 3 m and mass 12 kg, is pivoted at C, where $AC = 1$ m. A vertical force **F** applied at A maintains the rod in horizontal equilibrium. Calculate the magnitude of **F**.

5 A broom consists of a broomstick of length 130 cm and mass 5 kg, and a broomhead of mass 5.5 kg attached at one end. By modelling the broomstick as a uniform rod and the broomhead as a particle, find where a support should be placed so that the broom will balance horizontally.

(P) 6 A uniform rod AB, of length 4 m and weight 20 N, is suspended horizontally by two vertical strings attached at A and at B. A particle of weight 10 N is attached to the rod at point C, where $AC = 1.5$ m.

   a Find the magnitudes of the tensions in the two strings.

   The particle is moved so that the rod remains in horizontal equilibrium with the tension in the string at B 1.5 times the tension in the string at A.

   b Find the new distance of the particle from A.

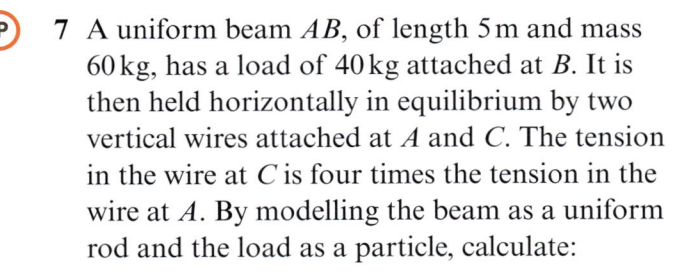

(E/P) 7 A uniform beam AB, of length 5 m and mass 60 kg, has a load of 40 kg attached at B. It is then held horizontally in equilibrium by two vertical wires attached at A and C. The tension in the wire at C is four times the tension in the wire at A. By modelling the beam as a uniform rod and the load as a particle, calculate:

   a the tension in the wire at C **(2 marks)**

   b the distance CB. **(5 marks)**

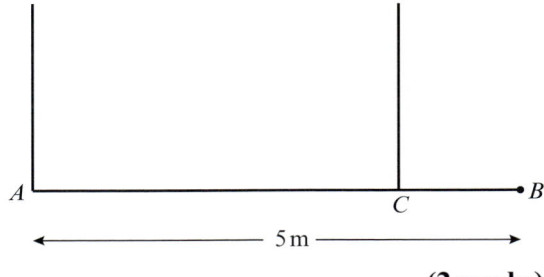

CHAPTER 8 — MOMENTS

8 A uniform plank AB has length $5\,\text{m}$ and mass $15\,\text{kg}$. The plank is held in equilibrium horizontally by two smooth supports A and C as shown in the diagram, where $BC = 2\,\text{m}$.

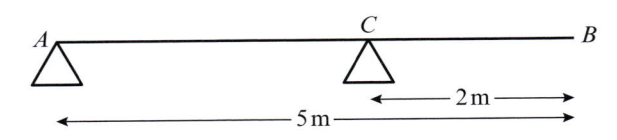

   a Find the reaction on the plank at C. **(3 marks)**

   A person of mass $45\,\text{kg}$ stands on the plank at the point D and it remains in equilibrium. The reactions on the plank at A and C are now equal.

   b Find the distance AD. **(7 marks)**

9 A uniform beam AB has weight $W\,\text{N}$ and length $8\,\text{m}$. The beam is held in a horizontal position in equilibrium by two vertical light inextensible wires attached to the beam at the points A and C where $AC = 4.5\,\text{m}$, as shown in the diagram. A particle of weight $30\,\text{N}$ is attached to the beam at B.

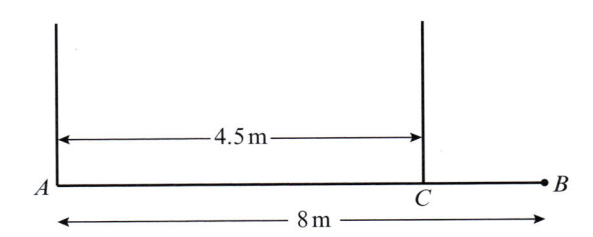

   a Show that the tension in the wire attached to the beam at C is $\left(\frac{8}{9}W + \frac{160}{3}\right)\text{N}$. **(4 marks)**

   b Find, in terms of W, the tension in the wire attached to the beam at A. **(3 marks)**

   c Given that the tension in the wire attached to the beam at C is twelve times the tension in the wire attached to the beam at A, find the value of W. **(3 marks)**

Challenge

The diagram shows a kinetic sculpture made from hanging rods. The distances between the points marked on each rod are equal. Arrange 1 kg, 2 kg, 3 kg, 4 kg and 5 kg weights onto the marked squares, using each weight once, so that the sculpture hangs in equilibrium with the rods horizontal.

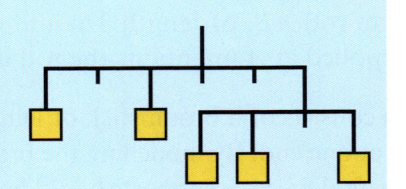

8.4 Centres of mass

So far you have considered only **uniform** rods, where the centre of mass is always at the midpoint. If a rod is **non-uniform** the centre of mass is not necessarily at the midpoint of the rod.

You might need to consider the moment due to the weight of a non-uniform rod, or find the position of its centre of mass.

Example 6 — SKILLS PROBLEM-SOLVING

Sam and Tamsin are sitting on a non-uniform plank AB of mass $25\,\text{kg}$ and length $4\,\text{m}$. The plank is pivoted at M, the midpoint of AB. The centre of mass of AB is at C where AC is $1.8\,\text{m}$. Sam has mass $35\,\text{kg}$. Tamsin has mass $25\,\text{kg}$ and sits at A.
Where must Sam sit for the plank to be horizontal?

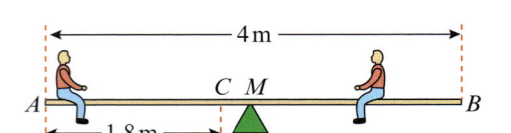

MOMENTS
CHAPTER 8

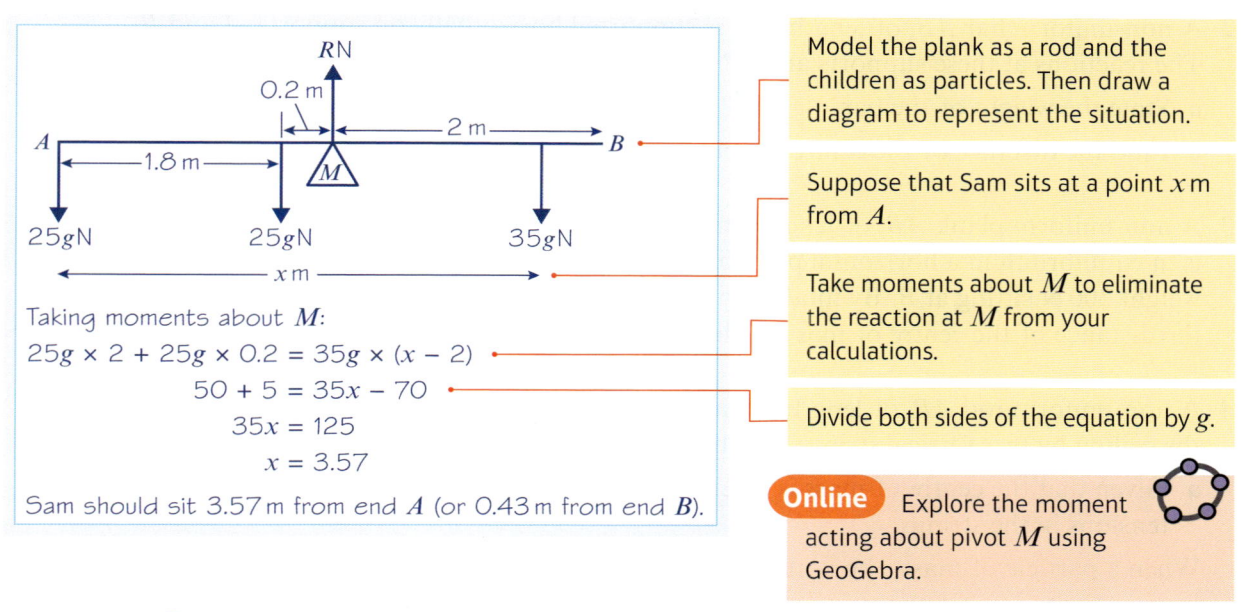

Taking moments about M:
$25g \times 2 + 25g \times 0.2 = 35g \times (x - 2)$
$50 + 5 = 35x - 70$
$35x = 125$
$x = 3.57$

Sam should sit 3.57 m from end A (or 0.43 m from end B).

Model the plank as a rod and the children as particles. Then draw a diagram to represent the situation.

Suppose that Sam sits at a point x m from A.

Take moments about M to eliminate the reaction at M from your calculations.

Divide both sides of the equation by g.

Online Explore the moment acting about pivot M using GeoGebra.

Example 7 — SKILLS PROBLEM-SOLVING

A non-uniform rod AB is 3 m long and has weight 20 N. It is in a horizontal position resting on supports at points C and D, where $AC = 1$ m and $AD = 2.5$ m. The magnitude of the reaction at C is three times the magnitude of the reaction at D. Find the distance of the centre of mass of the rod from A.

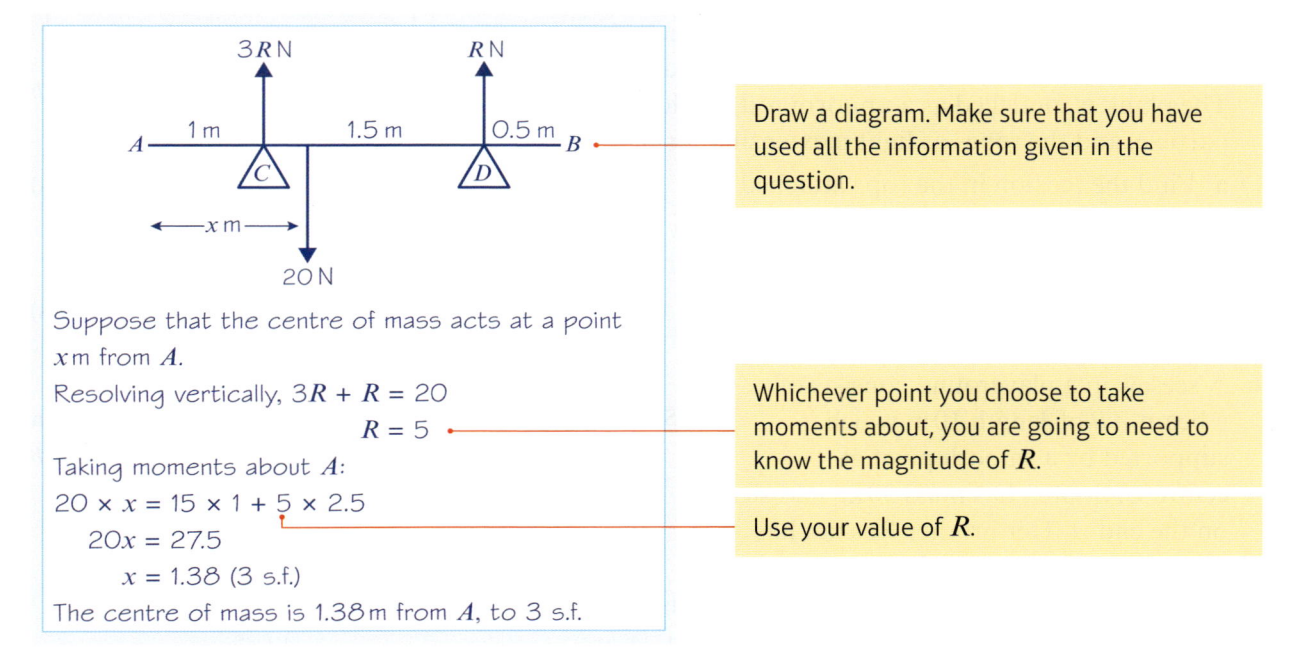

Suppose that the centre of mass acts at a point x m from A.
Resolving vertically, $3R + R = 20$
$R = 5$
Taking moments about A:
$20 \times x = 15 \times 1 + 5 \times 2.5$
$20x = 27.5$
$x = 1.38$ (3 s.f.)
The centre of mass is 1.38 m from A, to 3 s.f.

Draw a diagram. Make sure that you have used all the information given in the question.

Whichever point you choose to take moments about, you are going to need to know the magnitude of R.

Use your value of R.

Exercise 8D — SKILLS PROBLEM-SOLVING

1 A non-uniform rod AB, of length 4 m and weight 6 N, rests horizontally on two supports, A and B. Given that the centre of mass of the rod is 2.4 m from A, find the reactions at the two supports.

2 A non-uniform bar AB, of length 5 m, is supported horizontally on supports, A and B. The reactions at these supports are $3g$ N and $7g$ N respectively.

  a State the weight of the bar.

  b Find the distance of the centre of mass of the bar from A.

3 A non-uniform plank AB, of length 4 m and weight 120 N, is pivoted at its midpoint. The plank is in equilibrium in a horizontal position with a child of weight 200 N sitting at A and a child of weight 300 N sitting at B. By modelling the plank as a rod and the two children as particles, find the distance of the centre of mass of the plank from A.

(P) 4 A non-uniform rod AB, of length 5 m and mass 15 kg, rests horizontally suspended from the ceiling by two vertical strings attached to C and D, where $AC = 1$ m and $AD = 3.5$ m.

  a Given that the centre of mass is at E, where $AE = 3$ m, find the magnitudes of the tensions in the strings.

  When a particle of mass 9 kg is attached to the rod at F, the tension in the string at D is twice the tension in the string at C.

  b Find the distance AF.

(E/P) 5 A plank AB has mass 24 kg and length 4.8 m. A load of mass 15 kg is attached to the plank at the point C, where $AC = 1.4$ m. The loaded plank is held in equilibrium, with AB horizontal, by two vertical ropes, one attached at A and the other attached at B, as shown in the diagram. The plank is modelled as a uniform rod, the load as a particle, and the ropes as light inextensible strings.

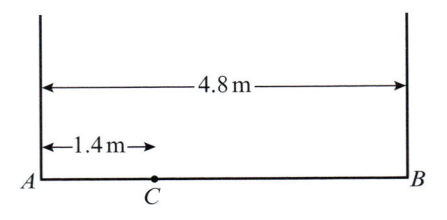

  a Find the tension in the rope attached at B. **(4 marks)**

  The plank is now modelled as a non-uniform rod. With the new model, the tension in the rope attached at A is 25 N greater than the tension in the rope attached at B.

  b Find the distance of the centre of mass of the plank from A. **(6 marks)**

(E) 6 A seesaw in a playground consists of a beam AB of length 10 m which is supported by a smooth pivot at its centre C. Sophia has mass 30 kg and sits on the end A. Her brother Roshan has mass 50 kg and sits at a distance x metres from C, as shown in the diagram. The beam is initially modelled as a uniform rod.

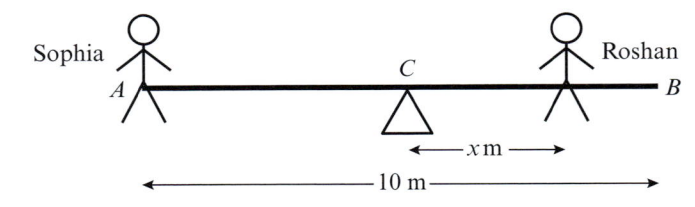

  a Using this model, find the value of x for which the seesaw can rest in equilibrium in a horizontal position. **(3 marks)**

  b State what is implied by the modelling assumption that the beam is uniform. **(1 mark)**

  Roshan finds he must sit at a distance 4 m from C for the seesaw to rest horizontally in equilibrium. The beam is now modelled as a non-uniform rod of mass 25 kg.

  c Using this model, find the distance of the centre of mass of the beam from C. **(4 marks)**

E/P 7 A non-uniform rod AB, of length 25 m and weight 80 N, rests horizontally in equilibrium on supports C and D as shown in the diagram. The centre of mass of the rod is 10 m from A.

A particle of weight W newtons is attached to the rod at a point E, where E is x metres from A. The rod remains in equilibrium and the magnitude of the reaction at C is five times the magnitude of the reaction at D.

Show that $W = \dfrac{400}{25 - 3x}$.

(5 marks)

8.5 Tilting

You need to be able to answer questions involving rods that are on the point of **tilting**.

- When a rigid body is on the point of tilting about a pivot, the reaction at any other support (or the tension in any other wire or string) is zero.

Example 8   SKILLS   PROBLEM-SOLVING

A uniform beam AB, of mass 45 kg and length 16 m, rests horizontally on supports C and D where $AC = 5$ m and $CD = 9$ m. When a child stands at A, the beam is on the point of tilting about C. Find the mass of the child.

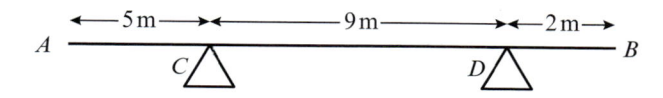

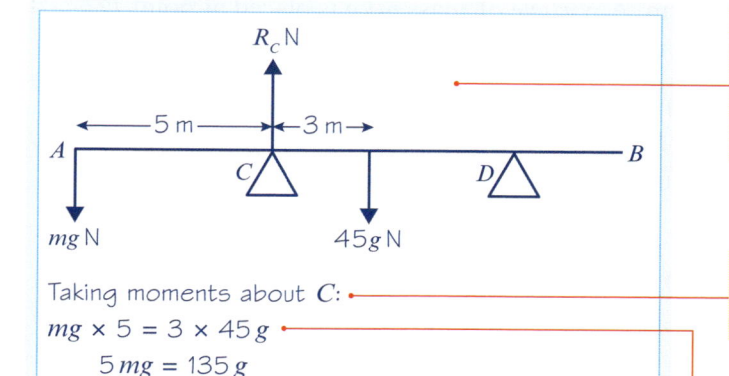

Draw a diagram showing the forces. Remember, as the beam is 'on the point of tilting', C is a pivot and the reaction force at D is zero.

Taking moments about C:

$mg \times 5 = 3 \times 45g$

$5mg = 135g$

$m = \dfrac{135g}{5g} = 27$

The mass of the child is 27 kg.

Take moments about C. This means you can ignore the reaction R_C.

Anticlockwise moment = Clockwise moment

Online See the point at which the beam starts to tilt due to the weight of the child and explore the problem with different forces and distances using GeoGebra.

Example 9 — SKILLS: PROBLEM-SOLVING

A non-uniform rod AB, of length 10 m and weight 40 N, is suspended from a pair of light cables attached to C and D where $AC = 3$ m and $BD = 2$ m.

When a weight of 25 N is hung from A the rod is on the point of rotating.

Find the distance of the centre of mass of the rod from A.

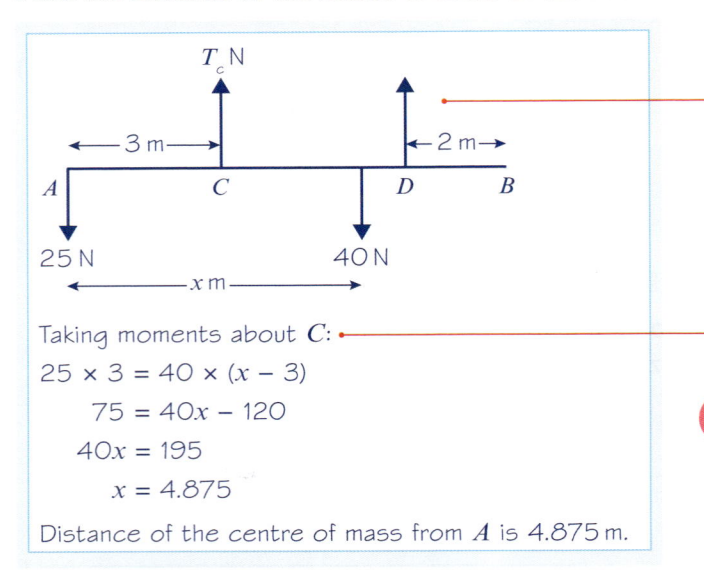

```
              T_c N
               ↑                    ↑
      ←─3 m─→              ←2 m→
   A─────────C──────────────D──────B
   ↓                        ↓
  25 N                     40 N
      ←────── x m ──────→
```

Draw a diagram showing the forces about C. As the rod is 'on the point of rotating', C is a pivot and the tension force at D is zero. As the rod is non-uniform you don't know the distance from A to the centre of mass, so you can label it as x.

Taking moments about C:

$25 \times 3 = 40 \times (x - 3)$

$75 = 40x - 120$

$40x = 195$

$x = 4.875$

Distance of the centre of mass from A is 4.875 m.

You don't require the tension force, T_C, so take moments about C.

Problem-solving

The distance from C to the centre of mass is $(x - 3)$ m. Equate the clockwise and anticlockwise moments then solve the equation to find the value of x.

Exercise 8E — SKILLS: PROBLEM-SOLVING

1 A uniform rod AB has length 4 m and mass 8 kg. It is resting in a horizontal position on supports at points C and D where $AC = 1$ m and $AD = 2.5$ m. A particle of mass m kg is placed at point E where $AE = 3.3$ m. Given that the rod is about to tilt about D, calculate the value of m.

2 A uniform bar AB, of length 6 m and weight 40 N, is resting in a horizontal position on supports at points C and D where $AC = 2$ m and $AD = 5$ m. When a particle of weight 30 N is attached to the bar at point E, the bar is on the point of tilting about C. Calculate the distance AE.

3 A plank AB, of mass 12 kg and length 3 m, is in equilibrium in a horizontal position resting on supports at C and D where $AC = 0.7$ m and $DB = 1.1$ m. A boy of mass 32 kg stands on the plank at point E. The plank is about to tilt about D. By modelling the plank as a uniform rod and the boy as a particle, calculate the distance AE.

(P) 4 A uniform rod AB has length 5 m and weight 20 N. The rod is resting on supports at points C and D where $AC = 2$ m and $BD = 1$ m.

   a Find the magnitudes of the reactions at C and D.

   A particle of weight 12 N is placed on the rod at point A.

   b Show that this causes the rod to tilt about C.

   A second particle of weight 100 N is placed on the rod at E to hold it in equilibrium.

   c Find the minimum and maximum possible distances of E from A.

5 A uniform plank of mass 100 kg and length 10 m rests horizontally on two smooth supports, A and B, as shown in the diagram. A man of mass 80 kg starts walking from one end of the plank, A, to the other end.

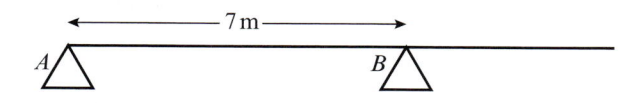

Find the distance he can walk past B before the plank starts to tip. **(4 marks)**

6 A non-uniform beam PQ, of mass m and length $8a$, hangs horizontally in equilibrium from two wires at M and N, where $PM = a$ and $QN = 2a$, as shown in the diagram. The centre of mass of the beam is at the point O. A particle of mass $\frac{3}{4}m$ is placed on the beam at Q and the beam is on the point of tipping about N.

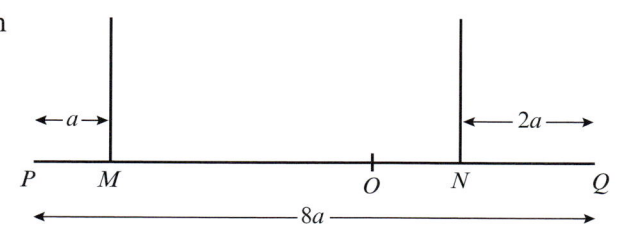

a Show that $ON = \frac{3}{2}a$. **(3 marks)**

The particle is removed and replaced at the midpoint of the beam and the beam remains in equilibrium.

b Find the magnitude of the tension in the wire attached at point N in terms of g. **(5 marks)**

7 A uniform beam AB, of weight WB and length 14 m, hangs in equilibrium in a horizontal position from two vertical cables attached at points C and D where $AC = 4$ m and $BD = 6$ m.

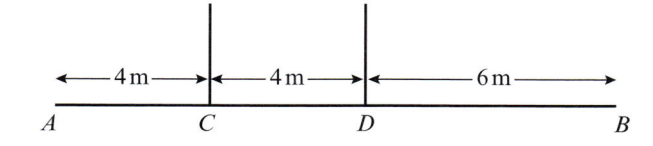

A weight of 180 N is hung from A and the beam is about to tilt. The weight is removed and a different weight, V N, is hung from B and the beam does not tilt. Find the maximum value of V. **(6 marks)**

Chapter review 8

1 A plank AE, of length 6 m and weight 100 N, rests in a horizontal position on supports at B and D, where $AB = 1$ m and $DE = 1.5$ m. A child of weight 145 N stands at C, the midpoint of AE, as shown in the diagram. The child is modelled as a particle and the plank as a uniform rod. The child and the plank are in equilibrium. Calculate:

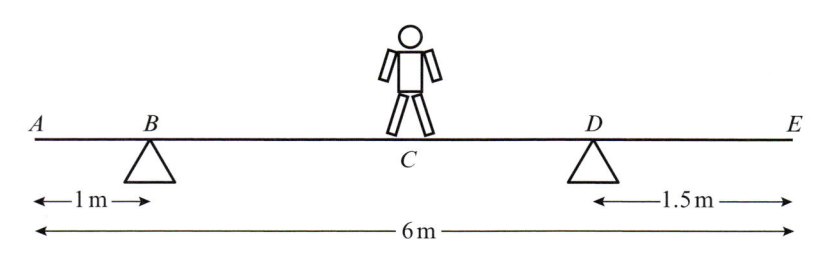

a the magnitude of the force exerted by the support on the plank at B **(3 marks)**

b the magnitude of the force exerted by the support on the plank at D. **(2 marks)**

The child now stands at a different point F on the plank. The plank is in equilibrium and on the point of tilting about D.

c Calculate the distance DF. **(4 marks)**

2 A uniform rod AB has length 4 m and weight 150 N. The rod rests in equilibrium in a horizontal position, smoothly supported at points C and D, where $AC = 1$ m and $AD = 2.5$ m, as shown in the diagram. A particle of weight W N is attached to the rod at point E where $AE = x$ metres. The rod remains in equilibrium and the magnitude of the reaction at C is now equal to the magnitude of the reaction at D.

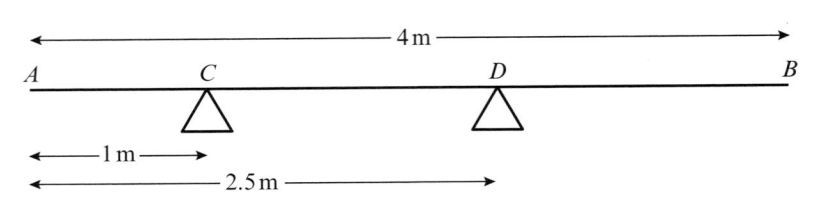

  a Show that $W = \dfrac{150}{7 - 4x}$. **(6 marks)**

  b Hence, deduce the range of possible values of x. **(3 marks)**

3 A uniform plank AB has mass 40 kg and length 4 m. It is supported in a horizontal position by two smooth pivots. One pivot is at the end A and the other is at the point C where $AC = 3$ m, as shown in the diagram. A man of mass 80 kg stands on the plank which remains in equilibrium. The magnitude of the reaction at A is twice the magnitude of the reaction at C. The magnitude of the reaction at C is R N. The plank is modelled as a rod and the man is modelled as a particle.

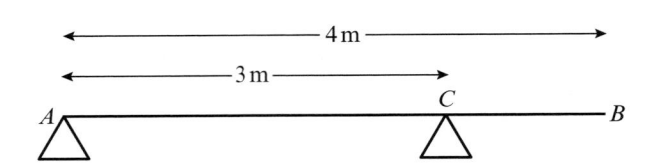

  a Find the value of R. **(2 marks)**

  b Find the distance of the man from A. **(3 marks)**

  c State how you have used the modelling assumption that:

    i the plank is uniform

    ii the plank is a rod

    iii the man is a particle. **(3 marks)**

4 A non-uniform rod AB has length 4 m and weight 150 N. The rod rests horizontally in equilibrium on two smooth supports C and D, where $AC = 1$ m and $DB = 0.5$ m, as shown in the diagram. The centre of mass of AB is x metres from A. A particle of weight W N is placed on the rod at A. The rod remains in equilibrium and the magnitude of the reaction of C on the rod is 100 N.

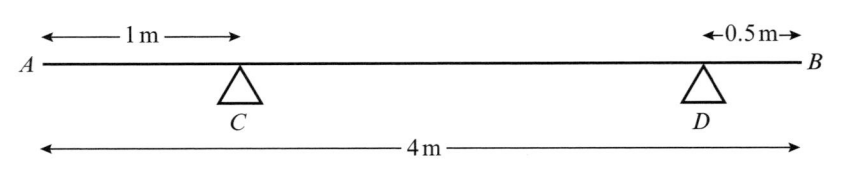

  a Show that $550 + 7W = 300x$. **(4 marks)**

The particle is now removed from A and placed on the rod at B. The rod remains in equilibrium and the reaction of C on the rod now has magnitude 52 N.

  b Obtain another equation connecting W and x. **(4 marks)**

  c Calculate the value of x and the value of W. **(3 marks)**

MOMENTS
CHAPTER 8

E 5 A lever consists of a uniform steel rod AB, of weight 100 N and length 2 m, which rests on a small smooth pivot at a point C. A load of weight 1700 N is suspended from the end B of the rod by a rope. The lever is held in equilibrium in a horizontal position by a vertical force applied at the end A, as shown in the diagram. The rope is modelled as a light string.

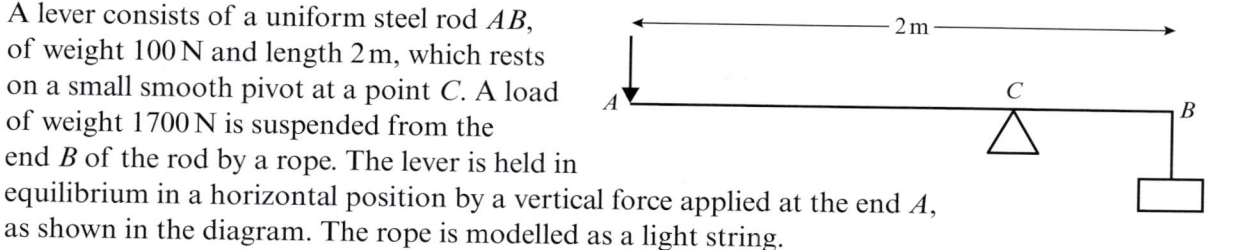

  a Given that $BC = 0.25$ m, find the magnitude of the force applied at A. **(4 marks)**

  The position of the pivot is changed so that the rod remains in equilibrium when the force at A has magnitude 150 N.

  b Find, to the nearest centimetre, the new distance of the pivot from B. **(4 marks)**

E 6 A plank AB has length 4 m. It lies on a horizontal platform, with the end A lying on the platform and the end B projecting over the edge, as shown in the diagram. The edge of the platform is at the point C.

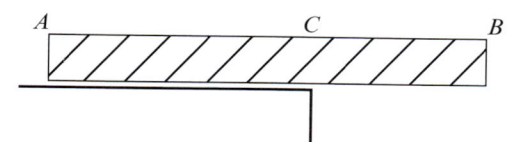

  Jack and Jill are experimenting with the plank. Jack has mass 48 kg and Jill has mass 36 kg. They discover that if Jack stands at B, Jill stands at A, and $BC = 1.8$ m, the plank is in equilibrium and on the point of tilting about C.

  a By modelling the plank as a uniform rod, and Jack and Jill as particles, find the mass of the plank. **(4 marks)**

  They now alter the position of the plank in relation to the platform so that, when Jill stands at B and Jack stands at A, the plank is again in equilibrium and on the point of tilting about C.

  b Find the distance BC in this position. **(4 marks)**

E 7 A plank of wood AB has mass 12 kg and length 5 m. It rests in a horizontal position on two smooth supports. One support is at the end A. The other is at the point C, 0.5 m from B, as shown in the diagram. A girl of mass 30 kg stands at B with the plank in equilibrium.

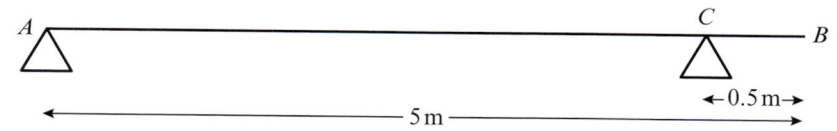

  a By modelling the plank as a uniform rod and the girl as a particle, find the reaction on the plank at A. **(4 marks)**

  The girl gets off the plank. A boulder of mass m kg is placed on the plank at A and a man of mass 93 kg stands on the plank at B. The plank remains in equilibrium and is on the point of tilting about C.

  b By modelling the plank again as a uniform rod, and the man and the boulder as particles, find the value of m. **(5 marks)**

E/P 8 A plank AB has mass 50 kg and length 4 m. A load of mass 25 kg is attached to the plank at B. The loaded plank is held in equilibrium, with AB horizontal, by two vertical ropes attached at A and C, as shown in the diagram. The plank is modelled as a uniform rod and the load as a particle. Given that the tension in the rope at C is four times the tension in the rope at A, calculate the distance CB. **(7 marks)**

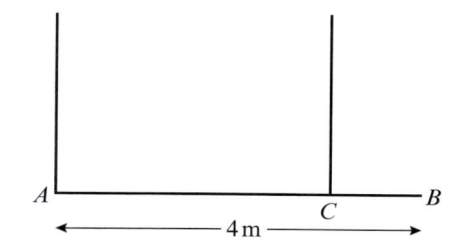

9 A beam AB has weight 200 N and length 5 m. The beam rests in equilibrium in a horizontal position on two smooth supports.

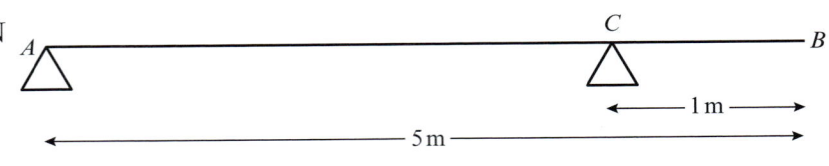

One support is at end A and the other is at a point C on the beam, where $BC = 1$ m, as shown in the diagram. The beam is modelled as a uniform rod.

a Find the reaction on the beam at C. **(4 marks)**

A woman of weight 500 N stands on the beam at the point D. The beam remains in equilibrium. The reactions on the beam at A and C are now equal.

b Find the distance AD. **(5 marks)**

10 The beam of a crane is modelled as a uniform rod AB, of length 30 m and weight 4000 kg, resting in horizontal equilibrium. The beam is supported by a tower at C, where $AC = 10$ m. A counterbalance mass of weight 3000 kg is placed at A and a load of mass M is placed a variable distance x m from the supporting tower, where $x \geqslant 5$.

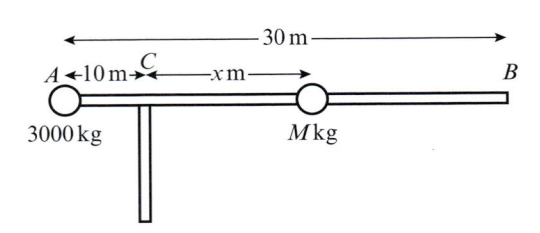

a Find an expression for M in terms of x. **(4 marks)**

b Hence, determine the maximum and minimum loads that can be lifted by the crane. **(2 marks)**

c Criticise this model in relation to the beam. **(1 mark)**

Challenge

1 A builder is attempting to tip over a refrigerator. The refrigerator is modelled as a rectangular lamina of weight 1200 N. The centre of mass of the lamina is at the point of intersection of the diagonals of the rectangle, as shown in the diagram.

Given that the refrigerator is pivoting at vertex P and that the base of the refrigerator makes an angle of 20° to the floor, find the minimum force needed to tip the refrigerator if the force is applied:

a horizontally at A

b vertically at B.

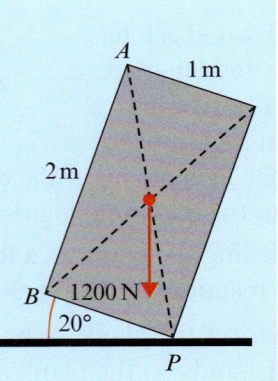

Summary of key points

1. Clockwise moment of **F** about $P = |\mathbf{F}| \times d$

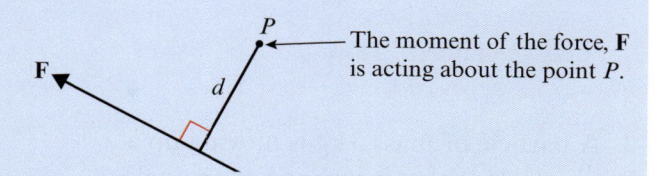

2. Clockwise moment of **F** about $P = |\mathbf{F}| \times d \sin \theta$

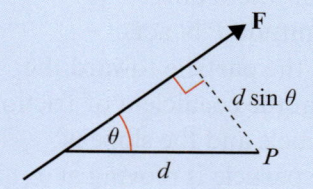

3. The sum of the moments acting on a body is called the resultant moment.

4. When a rigid body is in equilibrium, the resultant force in any direction is 0 N and the resultant moment about any point is 0 N m.

5. When a rigid body is on the point of tilting about a pivot, the reaction at any other support (or the tension in any other wire or string) is zero.

Review exercise 2

1. A particle of mass 3 kg is moving up a rough slope that is inclined at an angle α to the horizontal where $\tan \alpha = \frac{5}{12}$. A force of magnitude P N acts horizontally on the particle toward the plane. Given that the coefficient of friction between the particle and the slope is 0.2 and that the particle is moving at a constant velocity, calculate the value of P.
   ← Mechanics 1 Sections 5.2, 5.3

(P) 2. A particle of mass 2 kg sits on a smooth slope that is inclined at 45° to the horizontal. A force of F N acts at an angle of 30° to the plane on the particle causing it to accelerate up the hill at 2 m s^{-2}.

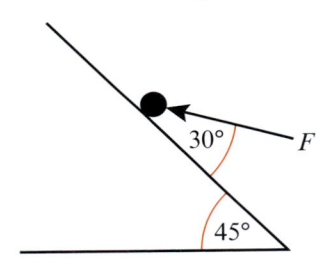

   Show that $F = 0.6$ N (2 s.f.)
   ← Mechanics 1 Sections 5.2, 5.3

(E/P) 3. A shipping container of mass 15 000 kg is being pulled by a winch up a rough slope that is inclined at 10° to the horizontal. The winch line imparts a constant force of 42 000 N, which acts parallel to and up the slope, causing the shipping container to accelerate at a constant rate of 0.1 m s^{-2}. Calculate:
   a. the reaction between the shipping container and the slope (2)
   b. the coefficient of friction, μ, between the shipping container and the slope. (3)

   When the shipping container is travelling at 2 m s^{-1} the engine is turned off.
   c. Find the time taken for the shipping container to come to rest. (3)
   d. Determine whether the shipping container will remain at rest, justifying your answer carefully. (2)
   ← Mechanics 1 Sections 5.2, 5.3

(E/P) 4. A ball of mass 0.3 kg is released at rest from a point at a height of 10 m above horizontal ground. After hitting the ground the ball rebounds to a height of 2.5 m.
   Calculate the magnitude of the impulse exerted by the ground on the ball. (4)
   ← Mechanics 1 Section 6.1

(E/P) 5. A toy racing car of mass 250 g is travelling on a smooth horizontal surface with momentum 2 N s. At point A it passes onto a rough horizontal surface where the coefficient of friction $\mu = 0.2$.
   a. Modelling the car as a particle which slides over the surface, find, to the nearest metre, the distance taken for the racing car to come to a complete stop. (6)
   b. In reality the car stops in a shorter distance. Suggest one reason for this. (1)
   ← Mechanics 1 Section 6.1

(E) 6. Two particles A and B have masses 0.4 kg and 0.3 kg respectively. They are moving in opposite directions on a smooth horizontal table and collide directly. Immediately before the collision, the speed of A is 6 m s^{-1} and the speed of B is 2 m s^{-1}. As a result of the collision, the direction of motion of B is reversed and its speed immediately after the collision is 3 m s^{-1}.

Find:
a the speed of A immediately after the collision, stating clearly whether the direction of motion of A is changed by the collision (3)
b the magnitude of the impulse exerted on B in the collision, stating clearly the units in which your answer is given. (3)

← Mechanics 1 Sections 6.1, 6.2

(E/P) 7 A railway truck S of mass 2000 kg is travelling due east along a straight horizontal track with constant speed 12 m s^{-1}. The truck S collides with a truck T which is travelling due west along the same track as S with constant speed 6 m s^{-1}. The magnitude of the impulse of T on S is 28 800 N s.

a Calculate the speed of S immediately after the collision. (2)
b State whether or not the motion of S is changed by the collision. (1)
c Given that, immediately after the collision, the speed of T is 3.6 m s^{-1}, and that T and S are moving in opposite directions, calculate the mass of T. (3)

← Mechanics 1 Sections 6.1, 6.2

(E/P) 8 Two particles A and B, of masses 0.5 kg and 0.4 kg respectively, are travelling in the same straight line on a smooth horizontal table. Particle A, moving with speed 3 m s^{-1}, strikes particle B, which is moving with speed 2 m s^{-1} in the same direction. After the collision A and B are moving in the same direction and the speed of B is 0.8 m s^{-1} greater than the speed of A.

a Find the speed of A and the speed of B after the collision. (5)
b Show that A loses momentum 0.4 N s in the collision. (3)

Particle B later hits an obstacle on the table and rebounds in the opposite direction with speed 1 m s^{-1}.

c Find the magnitude of the impulse received by B in this second impact. (3)

← Mechanics 1 Sections 6.1, 6.2

(E/P) 9 Two particles A and B, of masses 3 kg and 2 kg respectively, are moving in the same direction on a smooth horizontal table when they collide directly. Immediately before the collision, the speed of A is 4 m s^{-1} and the speed of B is 1.5 m s^{-1}. In the collision, the particles join to form a single particle C.

Find the speed of C immediately after the collision. (3)

← Mechanics 1 Sections 6.1, 6.2

(E/P) 10 Two particles P and Q have masses 3 kg and m kg respectively. They are moving toward each other in opposite directions on a smooth horizontal table. Each particle has speed 4 m s^{-1}, when they collide directly. In this collision, the direction of motion of each particle is reversed. The speed of P immediately after the collision is 2 m s^{-1} and the speed of Q is 1 m s^{-1}.

Find:
a the value of m (3)
b the magnitude of the impulse exerted on Q in the collision. (2)

← Mechanics 1 Sections 6.1, 6.2

(E/P) 11 A smooth bead B of mass 1 kg is threaded on a light inextensible string. The ends of the string are attached to two fixed points A and C where A is vertically above C. The bead is held in equilibrium by a horizontal force F. AB and BC make angles of 30° and 60° respectively with the vertical, as shown in diagram.

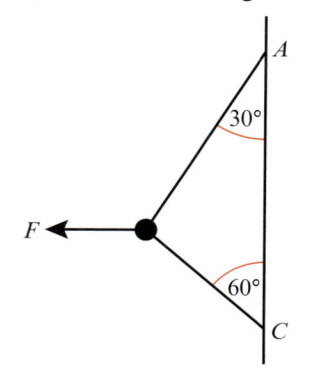

a Show that the tension in the string is $\frac{2g}{\sqrt{3}-1}$ N. (3)

b Calculate the magnitude of F. (3)

c State how you have used the fact that the bead is smooth in your calculations. (1)

← Mechanics 1 Section 7.2

E 12 A crate of mass 500 kg sits on a hill which is inclined at an angle α to the horizontal where $\tan \alpha = \frac{7}{24}$. The coefficient of friction between the hill and the crate is 0.15, and the crate is held at rest by a force of magnitude F N which acts parallel to and up the line of greatest slope of the hill. By modelling the crate as a particle,

a show that the normal reaction of the hill on the crate is $480g$ N (3)

b work out the minimum value of F. (3)

← Mechanics 1 Section 7.3

13 A box of 12 kg sits on a rough horizontal table. It is connected to a second box of mass 3 kg via a pulley by a light inextensible string.

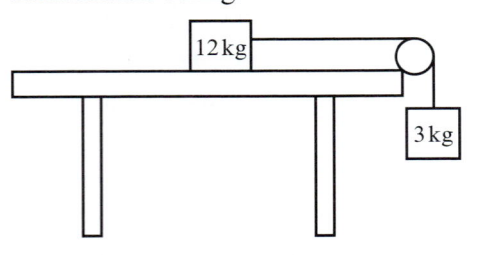

Given that the system is in limiting equilibrium, work out the value of the coefficient of friction of the table.

← Mechanics 1 Sections 7.2, 7.3

14 Two boxes of mass m_1 and m_2 sit on different rough slopes that are angled at θ_1 and θ_2 to the horizontal. The boxes are joined by a light inextensible string via a pulley as shown in the diagram.

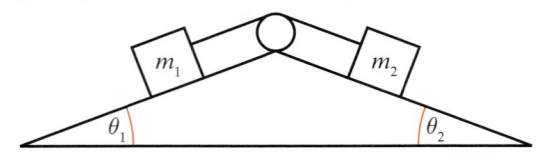

The coefficient of friction μ is the same for both slopes.

Given that the system is in limiting equilibrium,

show that $\mu = \dfrac{m_1 \sin\theta_1 - m_2 \sin\theta_2}{m_1 \cos\theta_1 - m_2 \cos\theta_2}$.

← Mechanics 1 Sections 7.2, 7.3

15 A box of mass 10 kg is released from rest from a point P that lies 10 m up a smooth plane that is inclined at 30° to the horizontal.

a Work out the speed of the box when it reaches the bottom of the slope.

At the bottom of the slope, the box meets a rough horizontal plane that has coefficient of friction $\mu = 0.2$.

b Work out how far the box travels from the instant it is released.

← Mechanics 1 Sections 7.2, 7.3

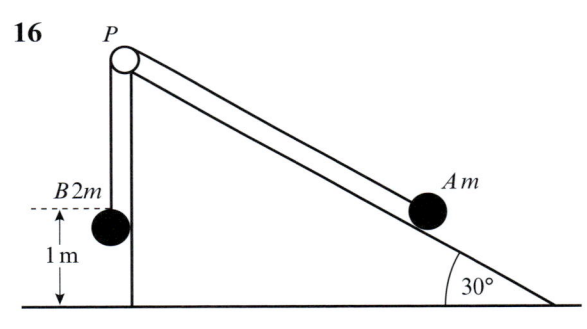

16

A box A of mass m kg is held at rest on a rough horizontal plane that is inclined at 30° to the horizontal. The coefficient of friction between the box and the plane is $\frac{1}{5}$. Box A is connected to a second box B of mass $2m$ kg by a light inextensible string that passes over a smooth pulley. The system is released from rest with the string taught and B at a height of 1 m above the ground. In the subsequent motion A does not hit the ground.

a Calculate the acceleration of the system when box A is released from rest. (9)

When B hits the ground, it does not rebound and comes to immediate rest.

b Find the distance travelled by A from the instant when the system is released until the instant when A first comes to rest. (7)

← **Mechanics 1 Section 7.3**

17 A non-uniform rod AB, of length 0.8 m, rests on the edge of a table as shown in the diagram.

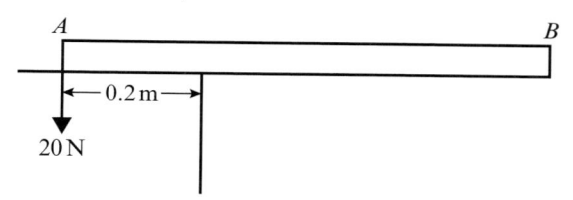

The centre of mass of the rod acts at a point 0.6 m from A and a force of 20 N is applied vertically downward at A. Given that the rod is on the point of tipping, show that the weight of the rod is 10 N.

← **Mechanics 1 Sections 8.4, 8.5**

Challenge

1 A lever consists of a uniform steel rod AC of weight 100 N and length $2k$ m, which rests on a pivot at B that has a height of $0.3k$ m. $AB = 0.5k$ m. A mass m kg is attached to the lever at A. The mass is lifted by means of a force of magnitude F N that is applied vertically downward at C. Show that $F > \frac{1}{3}(mg - 100)$ N.

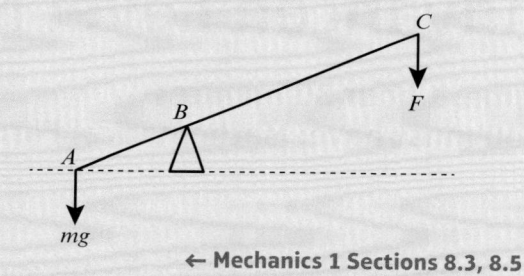

← **Mechanics 1 Sections 8.3, 8.5**

2 A sleigh is held at rest via a rope. The sleigh sits on a rough slope that is inclined at and angle of α to the horizontal. The coefficient of friction of the slope is μ. The rope is removed causing the sleigh to slide down the slope. Show that in t seconds the sleigh travels a distance of $\frac{1}{2}gt^2(\sin\alpha - \mu\cos\alpha)$.

← **Mechanics 1 Section 5.3**

3 Two identical balls A and B are moving toward each other on a smooth horizontal plane. Ball A has speed 4 m s^{-1} and ball B has speed 3 m s^{-1}. After the balls collide directly, the direction of motion of both balls is reversed. It then takes ball A 10 s to reach a point that is 5 m from the point of collision. Work out the speed of ball B after the collision.

← **Mechanics 1 Section 6.2**

Exam practice

Mathematics International Advanced Subsidiary/Advanced Level Mechanics 1

Time: 1 hour 30 minutes
You must have: Mathematical Formulae and Statistical Tables, Calculator

1 Two particles P and Q, of mass 4 kg and 2 kg respectively, move in the same direction on a smooth horizontal surface with constant speeds of $0.2\,\text{m\,s}^{-1}$ and $0.1\,\text{m\,s}^{-1}$ respectively. At $t = 0\,\text{s}$, P and Q are 0.8 m apart.

   a Work out the time taken for P and Q to collide. **(6)**

   After the collision, the speed of P is halved, and both P and Q continue to move in the same direction.

   b Work out the speed of Q. **(4)**

2 The velocity-time graph shows the motion of a train during a particular journey.

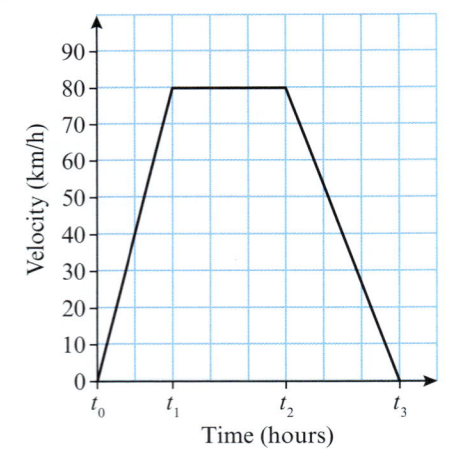

   a Describe what is happening between:
      i t_0 and t_1 **(1)**
      ii t_1 and t_2 **(1)**
      iii t_2 and t_3. **(1)**

   The train travels 120 km between t_1 and t_2. Work out:

   b the total length of the journey **(2)**

   c the total distance travelled by the train. **(4)**

3 A non-uniform rod AB of length $4\,m$ and of mass $20\,kg$ rests on two pivots P and Q that are $0.5\,m$ from A and B respectively. A mass of $50\,kg$ is placed at B causing the rod to be on the point of tipping around Q.

  a Work out the distance of the centre of mass from A. **(5)**

  The mass is removed from B.

  b Work out the reaction at P and Q. **(4)**

4 A particle P of mass $3m\,kg$ is attached by a light inextensible string to a particle Q of mass $m\,kg$ via a pulley. Particle Q is held at rest on the ground while particle P is at a height of $1.2\,m$ above the ground.

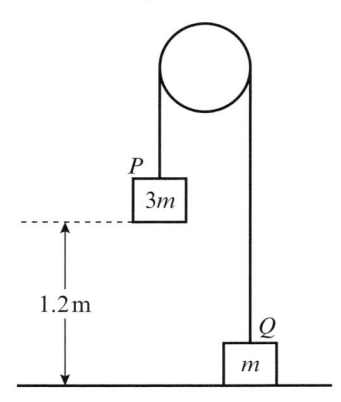

Particle Q is released.

  a Work out the acceleration of particle Q and the tension in the string. **(8)**

  b Show that the time taken for particle P to hit the ground is $0.353\,s$ (3 s.f.) **(4)**

5 A sledge A of mass $1200\,kg$ is connected to a second sledge B of mass $300\,kg$ by an inextensible tow-bar. The sledges sit on a rough horizontal track that has coefficient of friction μ. A forward force of $5940\,N$ is applied to sledge A causing it to accelerate at $2\,m\,s^{-2}$.

  a Show that $\mu = 0.2$. **(5)**

  b Work out the tension in the tow-bar. **(4)**

  c State one modelling assumption used in part **a**. **(1)**

  After 5 seconds the tow-bar snaps.

  d Assuming that resistance remains the same, work out how far sledge B travels before coming to rest. **(8)**

6 A ship leaves point A and sails with velocity $(5\mathbf{i} + 4\mathbf{j})\,\text{km}\,\text{h}^{-1}$. After 3 hours it reaches point B where it changes direction and sails with velocity $(8\mathbf{i} - 2\mathbf{j})\,\text{km}\,\text{h}^{-1}$. After 4 hours it reaches point C where it stops for the night.

  a Work out the average speed of the ship between A and C. **(5)**

  The next day the ship sails directly back to A at a speed of $10\,\text{km}\,\text{h}^{-1}$.

  b Work out the time taken for the ship to reach A. **(3)**

7 A particle A of mass m sits on a slope that is inclined at an angle of θ to the horizontal. The particle is held in equilibrium by two forces. One of the forces has magnitude P and acts horizontally on the particle. The other force has magnitude P and acts parallel to the slope.

  a Show that $P = \dfrac{mg\sin\theta}{1 + \cos\theta}$. **(3)**

  b Show that the normal reaction between the particle and the plane is mg. **(3)**

  The force acting horizontally is removed.

  c Given that $P = 0.25mg$ and $\theta = 30°$, find the initial acceleration of the particle. **(3)**

TOTAL FOR PAPER: 75 MARKS

GLOSSARY

acceleration positive rate of change of velocity

acceleration-time graph a graph of acceleration versus time

air resistance resistance due to air

assumption accepted as true

at rest stationary; not moving

bead a particle with a hole in it

body an object

buoyancy the upward force on a body that allows it to float or rise when submerged in a liquid (i.e. underneath the surface)

centre of mass the point through which the mass of an object is concentrated

coefficient of friction a measure of roughness (having a surface that is not level or smooth), usually denoted μ

column vector a vector written in the form $\begin{pmatrix} p \\ q \end{pmatrix}$

component a part

compression the force acting on an object that is being pushed

constant does not change; remains the same

coplanar in the same plane

deceleration negative rate of change of velocity

directed line segment a segment (part of a line) that has distance and direction

displacement distance in a particular direction

displacement-time graph a graph of displacement versus time

equation of motion the $F = ma$ equation

equilibrium having zero resultant force

force a push or a pull

force diagram a diagram showing the forces acting on a body

friction resistance due to the roughness of a surface (the quality of a surface that is not level or smooth) on which a body is moving

gravity the force between any object and the Earth

impulse change in momentum

impulse-momentum principle states that the change in momentum of an object is equal to the impulse applied to it

inclined sloped; at an angle

inextensible doesn't stretch

kinematics the branch of mathematics that deals with motion

kinematics formulae formulae that deal with kinematics; also known as suvat formulae

lamina a two-dimensional object whose thickness can be ignored

light inextensible string a connecting string for which you can assume negligible mass and fixed length

light object an object of negligible mass

light pulley a pulley of negligible mass

limiting value the value below which a system remains constant

magnitude size

mass the measure of how much matter (physical substance) is in an object

maximum the largest possible value

mechanics the branch of mathematics that deals with motion and the action of forces on objects

midpoint the point of a line that divides it in two equal parts

minimum the lowest possible value

model an attempt to describe a system using a set of variables and a set of equations that define relationships between the variables

moment measures the turning effect of force on a rigid body

momentum the mass of an object multiplied by its velocity

motion movement; the way something moves

negative less than zero

negligible very small or unimportant; not necessary to consider

newton unit of force

Newton's first law of motion states that an object at rest will stay at rest, and that an object moving with constant velocity will continue to move with constant velocity unless an unbalanced force acts on the object

Newton's second law of motion states that the force needed to accelerate a particle is equal to the product of the mass of the particle and the acceleration produced

GLOSSARY

Newton's third law of motion states that for every action there is an equal and opposite reaction

non-uniform a body with mass that is not evenly distributed

normal reaction the force exerted on an object by the surface on which it rests

opposite on the other side of something; the reverse of

parallel two lines side-by-side, the same distance apart at every point

parallelogram law when two vector quantities are represented by two adjacent sides of a parallelogram then the resultant of these vectors is represented by the diagonal of the parallelogram

particle an object with negligible dimensions

pivot the point on which an object (usually a rod) rests

plane a level surface

position vector describes the position of an object relative to (i.e. in relation to) a given point, usually the origin, using vector notation

positive greater than zero

principle of conservation of momentum states that in a collision (crash) between two objects, the total momentum of the two objects before the collision is equal to the total momentum of the two objects after the collision

rate of change the ratio between a change in one variable relative to (i.e. in relation to) a similar or related change in another variable

resistance opposition

resultant the sum of two or more quantities

rigid body a solid body which is unchanging or not easily changed

rod a thin straight object that does not bend

rough surface a surface that has friction

scalar a quantity with magnitude only

shunt a rigid, inextensible rod used to connect two bodies

SI units 'Système International d'Unités'; international units of measurement

smooth without friction

speed the measure of how quickly a body moves

speed of projection the measure of how quickly a body moves at the instant of release

static not moving

suvat formulae formulae that deal with motion; also known as kinematics formulae

taut stretched or pulled tight; not slack

tension the pulling force passed via a string or other continuous object

thrust the force acting on an object that is being pushed

tilt to position with one side or end higher than the other side

triangle law states that when two vectors (or forces) are represented by two sides of a triangle, then the third side of that triangle represents the resultant of the vectors (or forces)

uniform body a body with evenly distributed mass

uniformly not varying; the same in all parts and at all times

unit vector a vector with a magnitude of 1

validity how true or reasonable (valid) something is

vector a quantity that has both magnitude and direction

vector addition the addition of vector quantities

velocity the rate of change of displacement

velocity-time graph a graph of velocity versus time

weight the downward force caused by gravity on an object

ANSWERS

CHAPTER 1

Prior knowledge check
1. **a** $x = 4$ or $x = \frac{1}{5}$ **b** $x = \frac{3}{2}$ or $x = -\frac{7}{3}$
   c $x = 2.26$ or $x = -0.591$ **d** $x = \pm\frac{3}{2}$
2. **a** $x = 10.3, y = 61.0°$ **b** $x = 14.8, y = 8.7$
3. **a** $833\,\text{cm s}^{-1}$ **b** $5000\,\text{kg m}^{-3}$
4. **a** 7.65×10^6 **b** 3.806×10^{-3}

Exercise 1A
1. **a i** $h = 0$ **ii** $h = 6\,\text{m}$
   b $h = -48\,\text{m}$
   c Model is not valid when $x = 200$ as height would be $48\,\text{m}$ below ground level.
2. **a** $90\,\text{m}$
   b i $h = 90\,\text{m}$ **ii** $h = 40\,\text{m}$
   c $h = -1610\,\text{m}$
   d Model is not valid when $t = 20$ as height would be $1610\,\text{m}$ below sea level.
3. **a** $x = 2.30\,\text{m}$ or $8.70\,\text{m}$
   b $k = 10\,\text{m}$
   c When $k = 10$ metres the ball passes through the net so the model is not valid for $k > 10$.
4. **a** $1320\,\text{m}$
   b Model is valid for $0 \leq t \leq 10$.
5. $0 \leq x \leq 120$
6. $0 \leq t \leq 6$

Exercise 1B
1. **a** Ignore the rotational effect of any external forces that are acting on it, and the effects of air resistance.
   b Ignore the frictional effects on the football due to air resistance.
2. **a** Ignore the rotational effect of any external forces that are acting on it, and the effects of air resistance.
   b Ignore any friction between the ice hockey puck and the ice surface.
3. The parachutist and parachute should be considered together as one particle as they move together.
4. **a** If modelled as a light rod, the fishing rod is considered to have no thickness and is rigid.
   b If the fishing rod had no thickness and was rigid it would be unsuitable for fishing.
5. **a** Model golf ball as a particle; ignore the effects of air resistance.
   b Model child on sledge as a particle; consider the hill as smooth.
   c Model objects as particles; string as light and inextensible; pulley as smooth.
   d Model suitcase and handle as a particle; path as smooth; ignore friction.

Exercise 1C
1. **a** $18.1\,\text{m s}^{-1}$ **b** $150\,\text{kg m}^{-2}$ **c** $5 \times 10^{-3}\,\text{m s}^{-1}$
   d $0.024\,\text{kg m}^{-3}$ **e** $45\,\text{kg m}^{-3}$ **f** $63\,\text{kg m}^{-2}$
2. **a** A: Normal reaction, B: Forward thrust, C: Weight, D: Friction.
   b A: Buoyancy, B: Forward thrust, C: Weight, D: Water resistance or drag.
   c A: Normal reaction, B: Friction, C: Weight, D: Tension.
   d A: Normal reaction, B: Weight, C: Friction.

Chapter review 1
1. **a** $3.6\,\text{m}$
   b $1\,\text{m}$ and $7\,\text{m}$
   c $0 \leq x \leq 8$
   d $4.8\,\text{m}$
2. **a** $7.68\,\text{m}$
   b $4.15\,\text{m}$
   c Ignore the effects of air resistance on the diver and rotational effects of external forces.
   d Assumption not valid; diver experiences drag and buoyancy in the water.
3. **a** Model the man on skis as a particle – ignore the rotational effect of any forces that are acting on the man as well as any effects due to air resistance. Consider the snow-covered slope as smooth – assume there is no friction between the skis and the snow-covered slope.
   b Model the yo-yo as a particle – ignore the rotational effect of any forces that are acting on the yo-yo as well as any effects due to air resistance. Consider the string as light and inextensible – ignore the weight of the string and assume it does not stretch. Model the yo-yo as smooth – assume there is no friction between the yo-yo and the string.
4. **a** $41.7\,\text{m s}^{-1}$
   b $6000\,\text{kg m}^{-2}$
   c $1.2 \times 10^6\,\text{kg m}^{-3}$
5. **a** Model ball as a particle. Assume the floor is smooth.
   b i Positive – the positive direction is defined as the direction in which the ball is travelling.
      ii Negative – the ball will be slowing down.

CHAPTER 2

Prior knowledge check
1. **a i** 3 **ii** 73.5
   b i 2 **ii** 150
   c i -1.5 **ii** 26.25
2. 26.25 miles
3. **a** $x = 2, y = -1.5$
   b $x = 1.27$ or $x = -2.77$

Exercise 2A
1. **a** $A\ 80\,\text{km h}^{-1}$, $B\ 40\,\text{km h}^{-1}$, $C\ 0\,\text{km h}^{-1}$, $D\ \text{km h}^{-1}$, $E\ -66.7\,\text{km h}^{-1}$
   b $0\,\text{km h}^{-1}$ **c** $50\,\text{km h}^{-1}$
2. **a** $187.5\,\text{km}$ **b** $50\,\text{km h}^{-1}$
3. **a** $12\,\text{km h}^{-1}$ **b** $12{:}45$
   c $-10\,\text{km h}^{-1}, 3\,\text{km h}^{-1}$ **d** $7.5\,\text{km h}^{-1}$
4. **a** $2.5\,\text{m}, 0.75\,\text{s}$ **b** $0\,\text{m s}^{-1}$
   c i The velocity of the ball is positive (upward). The ball is decelerating until it reaches 0 at the highest point.
      ii The velocity of the ball is negative (downward), and the ball is accelerating.

Exercise 2B

1 **a** 2.25 m s^{-2} **b** 90 m

2 **a** $v \text{ (m s}^{-1})$ graph: rises to 10, constant until 30, decreases to 0 at 42, t (s)

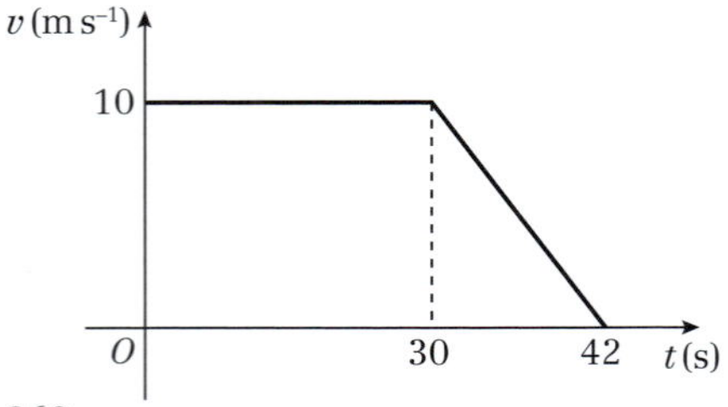

  b 360 m

3 **a** 0.4 m s^{-2} **b** $\frac{8}{15} \text{ m s}^{-2}$ or 0.53 m s^{-2} **c** 460 m

4 **a** $v \text{ (m s}^{-1})$ graph: trapezium reaching 30, with segments 15, T, 25, t(s)

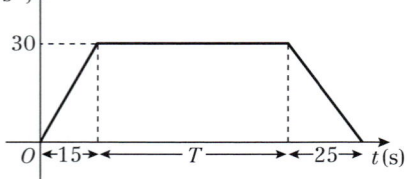

  b 100 s

5 **a** $v \text{ (m s}^{-1})$ graph: trapezium reaching 12, with segments 20, T, 40, t(s)

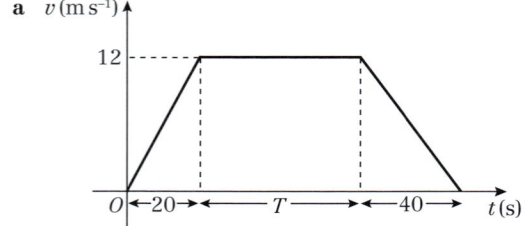

  b $T = 320$ **c** 3840 m

6 **a** $v \text{ (m s}^{-1})$ graph: trapezium reaching 10, with segments 15, $4T$, T, t(s)

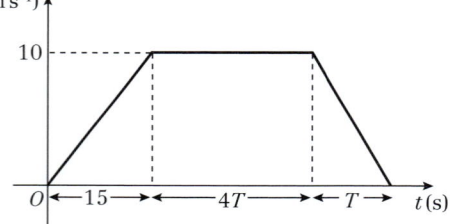

  b 60 s

7 **a** $u = \frac{10}{3}$ **b** $\frac{20}{9} \text{ m s}^{-2} = 2.22 \text{ m s}^{-2}$

8 **a** $v \text{ (m s}^{-1})$ graph: two lines C reaching 30 and M reaching 24, times 8, 20, T t(s)

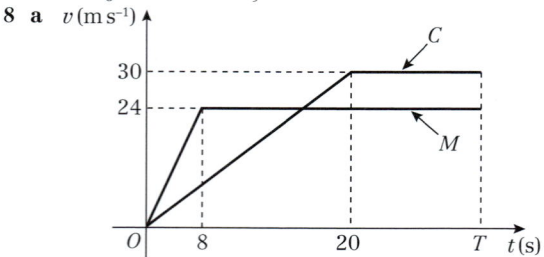

  b 720 m

Challenge

a 6 s **b** 16.5 m **c i** 10.5 m **ii** 4.5 m

Exercise 2C

1 **a** Slowing down (decelerating)

  b Constant velocity

  c Accelerating

  d 54 m s^{-1}

2 **a** a graph: level at 1 until 4, drops to 0 until 10, drops to -0.5 until 14, t

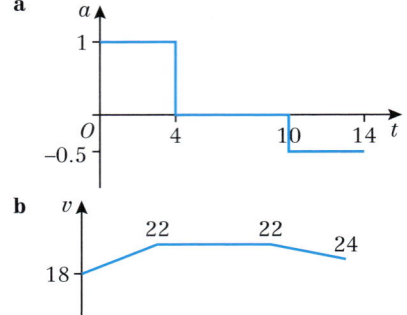

  b v graph: from 18 rising to 22 at 4, constant 22 until 10, down to 24 at 14, t

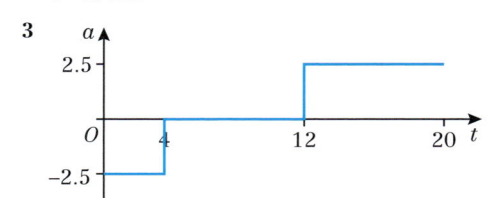

  c 296 m

3 a graph: -2.5 from 0 to 4, then 0 from 4 to 12, then 2.5 from 12 to 20, t

Challenge

a a graph: 5 from 0 to T, 0 from T to $T+6$, -4 from $T+6$ to $T+16$, t

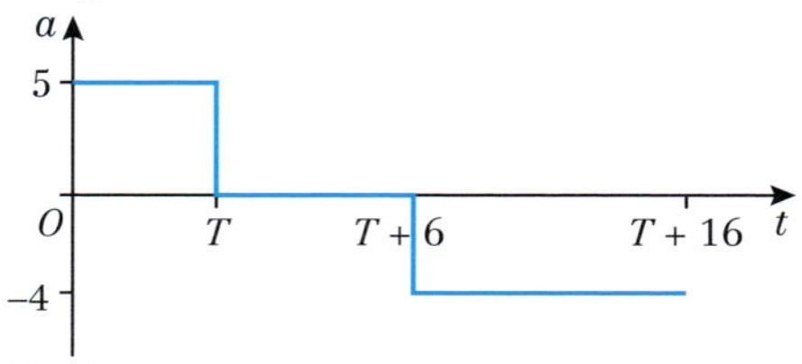

b $T = 4$

Exercise 2D

1 20 m s^{-1}

2 0.625 m s^{-2}

3 20 m s^{-1}

4 **a** 9 m s^{-1} **b** 72 m

5 **a** 3 m s^{-1} **b** $\frac{1}{3} \text{ m s}^{-2}$

6 **a** 9.2 m s^{-1} **b** 33.6 m

7 **a** 18 km h^{-1} **b** 312.5 m

8 **a** 8 s **b** 128 m

9 **a** 0.4 m s^{-2} **b** 320 m

10 **a** 0.25 m s^{-2} **b** 16 s **c** 234 m

11 **a** 19 m s^{-1} **b** 2.4 m s^{-2} **c** 430 m

12 **a** $x = 0.25$ **b** 150 m

13 **b** 500 m

Challenge

a $t = 3$ **b** 12 m

Exercise 2E

1 7 m s^{-1}

2 $\frac{2}{3} \text{ m s}^{-2}$

3 2 m s^{-2}

4 0.175 m s^{-2}

5 **a** 2.5 m s^{-2} **b** 4.8 s

6 **a** 3.5 m s^{-1} **b** 15.5 m s^{-1}

7 **a** 54 m **b** 6 s

8 **a** 90 m **b** 8.49 m s^{-1} (3 s.f.)

9 **a** 3.3 s (1 d.p.) **b** 16.2 m s^{-1} (1 d.p.)

10 a $t = 4$ or $t = 8$
   b $t = 4$: $4\,\text{m s}^{-1}$ in direction $\overrightarrow{AB}$,
     $t = 8$: $4\,\text{m s}^{-1}$ in direction $\overrightarrow{BA}$.
11 a $t = 0.8$ or $t = 4$
   b $15.0\,\text{m s}^{-1}$ (3 s.f.)
12 a $2\,\text{s}$ b $4\,\text{m}$
13 a $0.34\,\text{m s}^{-1}$ b $25.5\,\text{s}$ (3 s.f.)
14 a P: $(4t + t^2)\,\text{m}$ Q: $[3(t-1) + 1.8(t-1)^2]\,\text{m}$
   b $t = 6$ c $60\,\text{m}$
15 a $4.21\,\text{km h}^{-2}$ b $0.295\,\text{km h}^{-1}$

Exercise 2F

1 a $2.4\,\text{s}$ b $23.4\,\text{m s}^{-1}$
2 $4.1\,\text{s}$ (2 s.f.)
3 $41\,\text{m}$ (2 s.f.)
4 a $29\,\text{m}$ (2 s.f.)   b $2.4\,\text{s}$ (2 s.f.)
5 a $5.5\,\text{m s}^{-1}$ (2 s.f.)   b $20\,\text{m s}^{-1}$ (2 s.f.)
6 a $40\,\text{m s}^{-1}$ (2 s.f.)   b $3.7\,\text{s}$ (2 s.f.)
7 a $39\,\text{m s}^{-1}$ b $78\,\text{m}$ (2 s.f.)
8 $4.7\,\text{m}$ (2 s.f.)
9 a $3.4\,\text{s}$ (2 s.f.)   b $29\,\text{m}$ (2 s.f.)
10 $2.8\,\text{s}$ (2 s.f.)
11 a $u = 29$ (2 s.f.)   b $6\,\text{s}$
12 $30\,\text{m}$ (2 s.f.)
13 a $5.6\,\text{m}$ (2 s.f.)   b $3.2\,\text{m}$ (2 s.f.)

Challenge

1 a $1.4\,\text{s}$ (2 s.f.)   b $7.2\,\text{m}$ (2 s.f.)
2 $155\,\text{m}$ (3 s.f.)

Chapter review 2

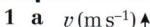

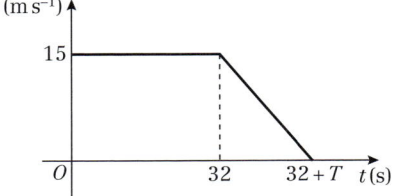

1 a *v-t graph: rises from O to 12.5 at t=20, horizontal to t=180*
   b $2125\,\text{m}$
2 a *v-t graph: horizontal at 15 until t=32, descends to 0 at t=32+T*
   b $T = 12$
   c *s-t graph: rises to 480 at t=32, then to 570 at t=44*

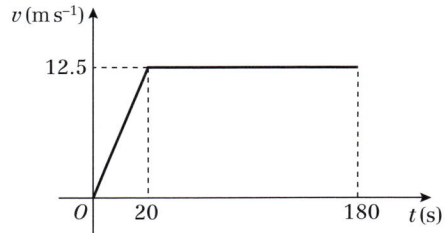

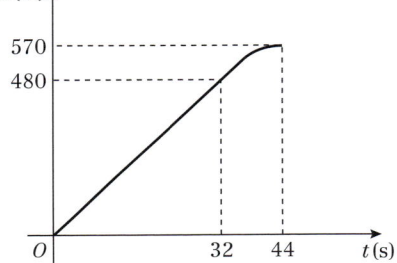

3 a i a = gradient of line. Using the formula for the gradient of a line, $a = \dfrac{v - u}{t}$, which can be rearranged to give $v = u + at$
     ii s = area under the graph. Using the formula for the area of a trapezium, $s = \left(\dfrac{u + v}{2}\right)t$
   b i Substitute $t = \dfrac{v - u}{a}$ into $s = \left(\dfrac{u + v}{2}\right)t$
     ii Substitute $v = u + at$ into $s = \left(\dfrac{u + v}{2}\right)t$
     iii Substitute $u = v - at$ into $s = \left(\dfrac{u + v}{2}\right)t$
4 $u = 8$
5 a *graph: a=1 for 0≤t<4, a=0 for 4≤t<8, a=−2 for 8≤t≤10*
6 $0.165\,\text{m s}^{-2}$ (3 d.p.)
7 a $60\,\text{m}$ b $100\,\text{m}$
8 $1.9\,\text{s}$
9 a i $4.1\,\text{s}$ (2 s.f.)   ii $40\,\text{m s}^{-1}$ (2 s.f.)
   b air resistance
10 a $u = 11$ b $22\,\text{m}$
11 a $28\,\text{m s}^{-1}$ b $208\,\text{m}$
12 a $8\,\text{m s}^{-1}$ b $1.25\,\text{m s}^{-2}$ c $204.8\,\text{m}$
13 a $33\,\text{m s}^{-1}$ (2 s.f.)   b $3.4\,\text{s}$ (2 s.f.)
   c *v-t graph: starts at 21, crosses t-axis at 2.1, reaches −33 at t=5.5*
14 a $50\,\text{s}$ b $24.2\,\text{m s}^{-1}$ (3 s.f.)
15 $h = 39$ (2 s.f.)
16 a $32\,\text{m s}^{-1}$ b $90\,\text{m}$ c $5\,\text{s}$
17 a *v-t graph: horizontal at 34 until t=2, descends to 22 at t=6, horizontal*
   b $180\,\text{m}$

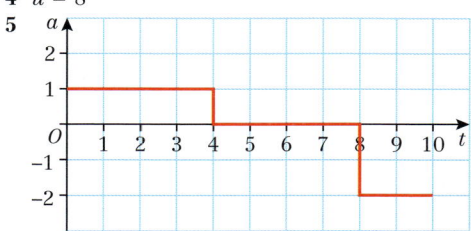

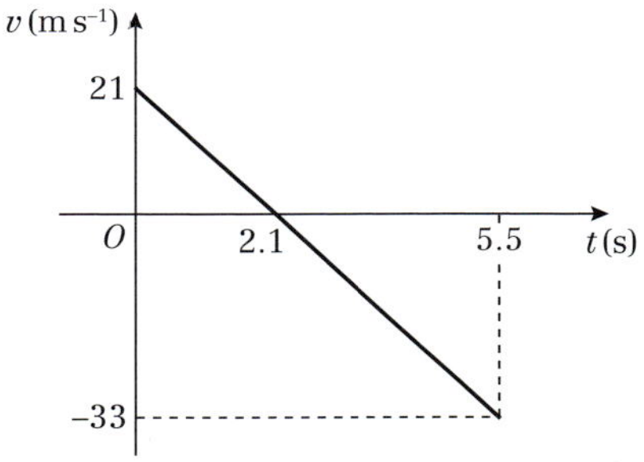

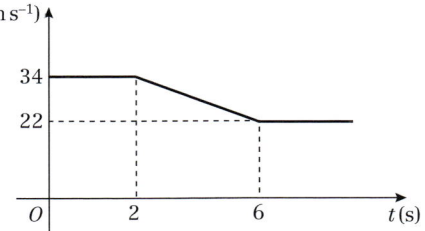

18 a

b $\frac{30}{t_1} = 3x \Rightarrow t_1 = \frac{1}{x}, \frac{-30}{t_2} = -x \Rightarrow t_2 = \frac{30}{x}$

So $\frac{10}{x} + T + \frac{30}{x} = 300 \Rightarrow \frac{40}{x} + T = 300$

c $T = 100, x = 0.2$ **d** 3 km   **e** 125 s

Challenge
1.2 s

CHAPTER 3
Prior knowledge check
1 **a** $\binom{4}{2}$ **b** $\binom{5}{-2}$ **c** $\binom{-1}{-3}$
2 **a** 123.2°   **b** 13.6   **c** 5.3   **d** 21.4°
3 **a** $\overrightarrow{AB} = -3\mathbf{a} + 4\mathbf{b}$
   b $|\overrightarrow{AB}| = 5$, Magnitude of $\overrightarrow{AB}$

Exercise 3A
1 8.60 km from the starting point on a bearing of 054°
2 10 km, 7.2 km on a bearing of 326°
3 **a** $v_1 = 4\mathbf{i}, v_2 = 5\mathbf{i} + 2\mathbf{j}, v_3 = -3\mathbf{i} + \mathbf{j},$
     $v_4 = 2\mathbf{i} + 3\mathbf{j}, v_5 = -2\mathbf{i} - \mathbf{j}, v_6 = -3\mathbf{j}$
   b **i** $9\mathbf{i} + 2\mathbf{j}$ **ii** $2\mathbf{j}$ **iii** $2\mathbf{i} - 4\mathbf{j}$
4 **a** 5   **b** 10   **c** 13
   d 4.47 (3 s.f.)   **e** 5.83 (3 s.f.)   **f** 8.06 (3 s.f.)
   g 5.83 (3 s.f.)   **h** 4.12 (3 s.f.)
5 **a** 5.10 (3 s.f.)   **b** 7.07 (3 s.f.)   **c** 14.0 (3 s.f.)
6 **a** 53.1° above   **b** 53.1° below
   c 67.4° above   **d** 63.4° above
7 **a** 149° to the right   **b** 29.7° to the right
   c 31.0° to the left   **d** 104° to the left

Challenge
Possible solution:

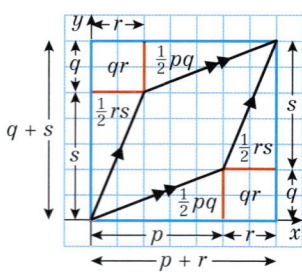

Area of parallelogram = area of large rectangle − 2(area of small rectangle) − 2 (area triangle 1) − 2(area triangle 2)
Area of parallelogram = $(p + r)(q + s) - 2qr - 2(\frac{1}{2}pq) - 2(\frac{1}{2}rs) = ps - qr$

Exercise 3B
1 **a** $6\mathbf{i} + 2\mathbf{j}$ **b** $10\mathbf{i} + 8\mathbf{j}$ **c** $7\mathbf{j}$ **d** $10\mathbf{i} + \mathbf{j}$
   e $-2\mathbf{i} + \mathbf{j}$ **f** $6\mathbf{i} + 2\mathbf{j}$ **g** $14\mathbf{i} - 7\mathbf{j}$ **h** $-8\mathbf{i} + 9\mathbf{j}$
2 **a** $\lambda = 5$ **b** $\mu = -\frac{3}{2}$
3 **a** $\lambda = \frac{1}{3}$ **b** $\mu = -1$ **c** $s = -1$ **d** $t = -\frac{1}{17}$
4 **a** 3.61, 023°   **b** 4.12, 104°
   c 3.61, 304°   **d** 2.24, 243°

Challenge
$\lambda = \frac{19}{5}$
In both cases, the direction and magnitude at both positions is the same, so in both positions λ is the same.

Exercise 3C
1 **a** $5\,\text{m s}^{-1}$ **b** $25\,\text{km h}^{-1}$
   c $5.39\,\text{m s}^{-1}$ **d** $8.06\,\text{cm s}^{-1}$
2 **a** 50 km   **b** 51.0 m   **c** 4.74 km   **d** 967 cm
3 **a** $5\,\text{m s}^{-1}$, 75 m   **b** $5.39\,\text{m s}^{-1}$, 16.2 m
   c $5.39\,\text{km h}^{-1}$, 16.2 km   **d** $13\,\text{km h}^{-1}$, 6.5 km

Exercise 3D
1 **a** $8\mathbf{i} + 3\mathbf{j}$ **b** $2\mathbf{i} - 7\mathbf{j}$ **c** $-17\mathbf{i} + 16\mathbf{j}$ **d** $7\mathbf{i} - 13\mathbf{j}$
2 **a** $2\mathbf{i} + 5\mathbf{j}$ **b** $\mathbf{i} + 3\mathbf{j}$ **c** $2\mathbf{i} + 4\mathbf{j}$
   d $2\mathbf{i} - 5\mathbf{j}$ **e** $-2\mathbf{i} - 5\mathbf{j}$
3 **a** $6\mathbf{i} - 8\mathbf{j}$ **b** $-12\mathbf{i} + 9\mathbf{j}$
   c $-4.5\mathbf{i} + 6\mathbf{j}$ **d** $5\mathbf{i} + 5\mathbf{j}$
   e $-4\mathbf{i} + 6\mathbf{j}$ **f** $3\sqrt{2}\mathbf{i} - 5\sqrt{2}\mathbf{j}$
   g $-4\sqrt{3}\mathbf{i} - 2\sqrt{3}\mathbf{j}$ **h** $-3\sqrt{5}\mathbf{i} + 6\sqrt{5}\mathbf{j}$
4 **a** $6\mathbf{i} + 12\mathbf{j}$ **b** $-7\mathbf{i} + 4\mathbf{j}$ **c** $-2\mathbf{i} + 6\mathbf{j}$ **d** $10\mathbf{i} - 13\mathbf{j}$
   e $2\mathbf{i} - 3\mathbf{j}$ **f** $4.61\,\text{m s}^{-1}$ **g** 4   **h** 2.5
5 **a** $5\mathbf{i} + 12\mathbf{j}, 13\,\text{m s}^{-1}$ **b** $6\mathbf{i} - 5\mathbf{j}, 7.81\,\text{m s}^{-1}$
   c $-2\mathbf{i} - 5\mathbf{j}, 5.39\,\text{m s}^{-1}$ **d** $-3\mathbf{i} - 2\mathbf{j}, 3.61\,\text{m s}^{-1}$
   e $7\mathbf{i} + 9\mathbf{j}, 11.4\,\text{m s}^{-1}$
6 $4.8\mathbf{i} - 6.4\mathbf{j}$
7 10.1 m
8 $2.03\,\text{m s}^{-1}$
9 **a** $2t\mathbf{i} + (-500 + 3t)\mathbf{j}$ **b** 721 m
10 **a** $7t\mathbf{i} + (400 + 7t)\mathbf{j}, (500 - 3t)\mathbf{i} + 15t\mathbf{j}$
   b $350\mathbf{i} + 750\mathbf{j}$
11 **a** $(1 + 2t)\mathbf{i} + (3 - t)\mathbf{j}, (5 - t)\mathbf{i} + (-2 + 4t)\mathbf{j}$
   b 5.39 km
12 **a** $121\,\text{m s}^{-1}, 6.08\,\text{m s}^{-1}$ **b** $18\mathbf{i} - 3\mathbf{j}$
   c $15\mathbf{i} - 12\mathbf{j}$

Challenge
24 seconds

Chapter review 3
1 **b** $(-3 + t)\mathbf{i} + (10 + t)\mathbf{j}$ **c** 4.24 km   **d** 1630
2 **a** $\mathbf{p} = 6t\mathbf{i}, \mathbf{q} = (12 - 3t)\mathbf{i} + (6 + 6t)\mathbf{j}$
   b 38.4 km
   c $1\frac{1}{3}$
3 **a** 3   **b** $10.2\,\text{m s}^{-1}$ **c** 168.7°
4 **a** $(-4\mathbf{i} + 2\mathbf{j})\,\text{m s}^{-2}$ **b** 22.4 N   **c** 26 m
5 **a** 031°   **b** $\mathbf{a} = 6t\mathbf{i}, \mathbf{b} = 3t\mathbf{i} + (-10 + 5t)\mathbf{j}$
   c 14:00   **e** 14:56
6 **a** 108°   **b** $(-2 + 9t)\mathbf{i} + (-4 - 3t)\mathbf{j}$ **c** 41, −23
7 **a** 124°   **b** $(3 - 2t)\mathbf{i} + (-2 + 3t)\mathbf{j}$
   c $11.2\,\text{m s}^{-1}$ **d** $t = 1$

8 a $9.85\,\text{m s}^{-1}$ b $(3 + 4t)\mathbf{i} + (2 + 9t)\mathbf{j}$
   c $6.5\,\text{s}$ d $7.46\,\text{m s}^{-1}$
9 a $(5\mathbf{i} + 3\mathbf{j})\,\text{km h}^{-1}$
   b $(10 + 5t)\mathbf{i} + (15 + 3t)\mathbf{j}$, $(-16 + 12t)\mathbf{i} + 26\mathbf{j}$
   d 05:25
10 a $2\sqrt{61}$ b 39.8° (3 s.f.)
11 a 5   b 143.1°

Challenge

$\overrightarrow{OB} = \frac{3}{2}\mathbf{i} + \frac{5}{2}\mathbf{j}$ or $\frac{99}{34}\mathbf{i} + \frac{5}{34}\mathbf{j}$

CHAPTER 4
Prior knowledge check
1 a $5\mathbf{i} - 3\mathbf{j}$ b $-4\mathbf{i} + 4\mathbf{j}$
2 a 19.2 cm   b 38.7°
3 a $18\,\text{m s}^{-1}$ b 162 m

Exercise 4A
1–5 (force diagrams)

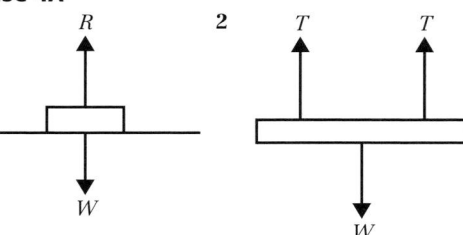

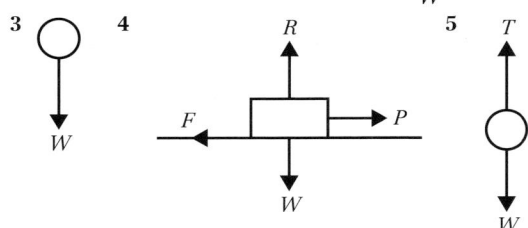

6 Although its speed is constant, the satellite is continuously changing direction. This means that the velocity changes. Therefore, there must be a resultant force on the satellite.
7 5 N
8 a 10 N   b 30 N   c 20 N
9 a 200 N
   b The platform accelerates toward the ground.
10 $p = 50, q = 40$
11 $P = 10\,\text{N}, Q = 5\,\text{N}$
12 a i 20 N   ii vertically upward
    b i 20 N   ii to the right
13 a (force diagram: R up, W down, 1600 N left, 10000 N right)
    b 8400 N
14 a (force diagram: R up, W down, F left, 8F right)
    b 600 N

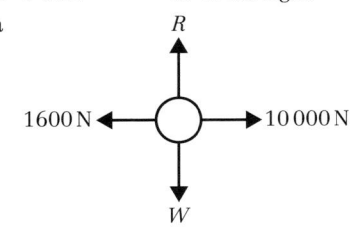

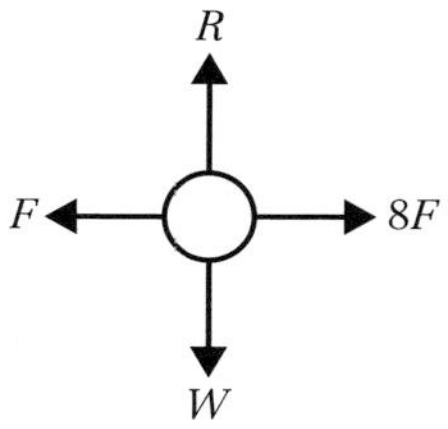

Exercise 4B
1 a $(3\mathbf{i} + 2\mathbf{j})\,\text{N}$ b $\begin{pmatrix} 2 \\ -3 \end{pmatrix}\,\text{N}$ c $(4\mathbf{i} - 3\mathbf{j})\,\text{N}$ d $\begin{pmatrix} 3 \\ -3 \end{pmatrix}\,\text{N}$
2 a $\mathbf{i} - 8\mathbf{j}$ b $-5\mathbf{i} + \mathbf{j}$
3 $a = 3, b = 4$
4 a i 5 N   ii 53.1°
   b i $\sqrt{26}\,\text{N}$ ii 11.3°
   c i $\sqrt{13}\,\text{N}$ ii 123.7°
   d i $\sqrt{2}\,\text{N}$ ii 135°
5 a i $(2\mathbf{i} - \mathbf{j})\,\text{N}$ ii $\sqrt{5}\,\text{N}$ iii 116.6°
   b i $(3\mathbf{i} + 4\mathbf{j})\,\text{N}$ ii 5 N   iii 036.9°
6 $a = 3, b = 1$
7 $a = 3, b = -1$
8 a $p = 2, q = -6$ b $\sqrt{40}\,\text{N}$ c 18°
9 a 63.4°   b 3.5
10 a $a = 3, b = 2$ b i $\sqrt{65}\,\text{N}$ ii 30°

Challenge
$a = 17.3$ (3 s.f.), magnitude of resultant force = 20 N

Exercise 4C
1 $0.3\,\text{m s}^{-2}$
2 39.2 N
3 25 kg
4 $1.6\,\text{m s}^{-2}$
5 a 25.6 N   b 41.2 N
6 a 2.1 kg (2 s.f.)   b 1.7 kg (2 s.f.)
7 a $5.8\,\text{m s}^{-2}$ b $2.7\,\text{m s}^{-2}$
8 4 N
9 a $0.9\,\text{m s}^{-2}$ b 7120 N   c 8560 N
10 a $0.5\,\text{m s}^{-2}$ b 45 N
11 a 32 s   b 256 m
    c The resistive force is unlikely to be constant.

Challenge
a 2.9 m (2 s.f.)   b $3.6\,\text{m s}^{-1}$ (2 s.f.)
c 2.16 s (3 s.f.)

Exercise 4D
1 a $(0.5\mathbf{i} + 2\mathbf{j})\,\text{m s}^{-2}$
   b $2.06\,\text{m s}^{-2}$ (3 s.f.) on a bearing of 014°
      (to the nearest degree).
2 0.2 kg
3 a $(21\mathbf{i} - 9\mathbf{j})\,\text{N}$
   b 22.8 N (3 s.f.) on a bearing of 113°
      (to the nearest degree).
4 a $(-4\mathbf{i} + 32\mathbf{j})\,\text{m s}^{-2}$ b $(\tfrac{5}{6}\mathbf{i} - \tfrac{1}{6}\mathbf{j})\,\text{m s}^{-2}$
   c $(-\mathbf{i} - \tfrac{2}{3}\mathbf{j})\,\text{m s}^{-2}$ d $(-\tfrac{4}{3}\mathbf{i} + 6\mathbf{j})\,\text{m s}^{-2}$
5 a $\sqrt{0.8125}\,\text{m s}^{-2}$ on a bearing of 146°
      (to the nearest degree).
   b 6.66 s
6 $\mathbf{R} = (-k\mathbf{i} + 4k\mathbf{j})\,\text{N}$
   So $4k = 3 + q$ (1), $-k = 2 + p$ (2) and $-4k = 8 + 4p$ (3)
   Adding equations (1) and (3) gives $4p + q + 11 = 0$

ANSWERS

7 a $b = 6$
   b $6\sqrt{2}$ N
   c $\dfrac{3\sqrt{2}}{2}$ m s^{-2}
   d $\dfrac{75\sqrt{2}}{4}$ m
8 a $p = 2, q = -6$
   b $\dfrac{25\sqrt{2}}{6}$ kg
9 0.86 kg
10 a $5 + q = -2k$ (1), $2 + p = k$ (2) and $4 + 2p = 2k$ (3)
   Adding equations (1) and (3) gives $2p + q + 9 = 0$
   b 0.2 kg

Challenge
$k = 8$

Exercise 4E

1 a 4 N **b** 0.8 N
   c Light ⇒ tension is the same throughout the length of the string and the mass of the string does not need to be considered. Inextensible ⇒ acceleration of masses is the same.
2 a 10 kg **b** 40 N
3 a 2 m s^{-2} **b** 14 N
4 a 16 000 N
   b i 880 N upward **ii** 2400 N downward
5 a 1800 kg and 5400 kg **b** 37 000 N
   c Light ⇒ tension is the same throughout the length of the tow-bar and the mass of the tow-bar does not need to be considered. Inextensible ⇒ acceleration of lorry and trailer is the same.
6 a 2.2 m s^{-2} **b** 60 N
7 a 4 kg **b** 47.2 N
8 a 6000 N
   b For engine, $F = ma = 3200$ N
   R(→) $12000 - 6000 - T = 3200$, $T = 2800$ N
9 a R(→) $1200 - 100 - 200 = 900$ N
   $F = ma$, so $a = 900 \div (300 + 900) = 0.75$ m s^{-2}
   b 325 N **c** 500 N

Exercise 4F

1 a 33.6 N (3 s.f.)
   b 2.37 m s^{-1} (3 s.f.)
   c 2.29 m (3 s.f.)
2 a $2mg$ N
   b For P: $2mg - kmg = \tfrac{1}{3}kmg$
   So $2 - k = \tfrac{1}{3}k$ and $k = 1.5$
   c Smooth ⇒ no friction so magnitude of acceleration is the same in objects connected by a taut inextensible string.
   d While Q is descending, distance travelled by $P = s_1$
   $s = ut + \tfrac{1}{2}at^2 \Rightarrow s_1 = \tfrac{1}{6}g \times 1.8^2 = 0.54g$
   Speed of P at this time $= v_1$
   $v^2 = u^2 + 2as \Rightarrow v_1^2 = 0^2 + \left(2 \times \dfrac{g}{3} \times 0.54g\right) = 0.36g^2$
   After Q hits the ground, P travels freely under gravity and travels a further distance s_2.
   $v^2 = u^2 + 2as \Rightarrow 0^2 = 0.36g^2 - 2gs_2 \Rightarrow s_2 = 0.18g$
   Total distance travelled $= s_1 + s_2 = 0.54g + 0.18g$
   $= 0.72g$ m
   As particles started at same height P must be s_1 metres above the plane at the start.
   Maximum height reached by P above the plane $= 0.72g + s_1 = 0.72g + 0.54g = 1.26g$ m
3 a $s = ut + \tfrac{1}{2}at^2$ so $2.5 = 0 + \tfrac{1}{2} \times a \times 1.25^2$, $a = 3.2$ m s^{-2}
   b 39 N
   c For A, R(↓): $mg - T = ma$
   $T = m(9.8 - 3.2)$, $T = 6.6m$
   Substituting for T: $39 = 6.6m$
   $m = \dfrac{65}{11}$
   d Same tension in string either side of the pulley.
   e $\dfrac{40}{49}$ s
4 a 0.613 m s^{-2} (3 s.f.)
   b 27.6 N (3 s.f.)
   c 39.0 N (3 s.f.)
5 a i 2.84 m s^{-2} (3 s.f.)
   ii $2.84(1.5) = 1.5g - T$
   $T = 1.5g - 4.26 = 10.4$ N (3 s.f.)
   iii 3.3 N
   b Same tension in string either side of the pulley.

Chapter review 4

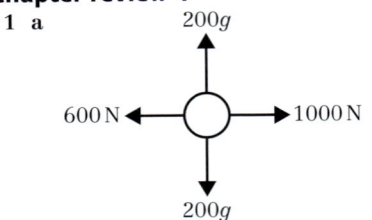

1 a

(Diagram: forces on object — 200g up, 200g down, 600 N left, 1000 N right)

   b 2 m s^{-2}
2 1000 N (2 s.f.) vertically downward
3 a 2000 N **b** 36 m
4 a 1.25 m s^{-2} **b** 6 N
5 Res(→) $3R - R = 1200 \times 2 \Rightarrow R = 1200$
   Driving force $= 3R = 3600$ N
6 $(28\mathbf{i} + 4\mathbf{j})$ m s^{-2}
7 $a = 1, b = -3$
8 a $\sqrt{5}$ m s^{-2} **b** $\dfrac{9\sqrt{5}}{2}$ m
9 a $a = -15, b = 12$
   b i 11.7 m s^{-2} (3 s.f.) on a bearing of 039.8° (3 s.f.)
   ii 52.7 m (3 s.f.)
10 a 0.7 m s^{-2}
   b 770 N
   c 58 m
   d Inextensible ⇒ tension the same throughout, and the acceleration of the car and the trailer is the same.
11 a R(→) $8000 - 500 - R = 3600 \times 1.75$, $R = 1200$ N
   b 2425 N
   c 630 N (2 s.f.)
12 a $\tfrac{1}{3}g$ m s^{-2}
   b 3.6 m s^{-1}
   c $2\tfrac{2}{3}$ m
   d i Acceleration both masses equal.
   ii Same tension in string either side of the pulley.
13 a $\tfrac{12}{7}g$ N **b** $m = 1.2$
14 a 3.2 m s^{-2}
   b 5.3 N (2 s.f.)
   c $F = 3.7$ (2 s.f.)
   d The information that the string is inextensible has been use in part **c** when the acceleration of A has been taken to be equal to the acceleration of B.
15 a i $0.5g - T = 0.5a$ **ii** $T - 0.4g = 0.4a$
   b $\tfrac{4}{9}g$ N **c** $\tfrac{1}{9}g$ m s^{-2} **d** 0.66 s (2 s.f.)

Challenge
$k = -\tfrac{5}{2}$

Review exercise 1

1. **a** Constant acceleration
   b Constant speed
   c 30.5 m
2. **a** $v\,(\text{m s}^{-1})$

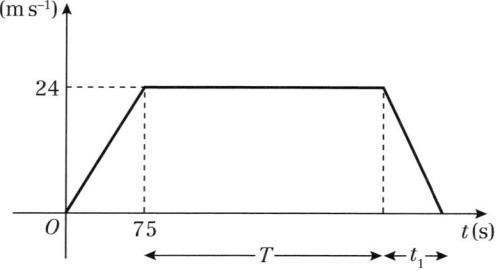

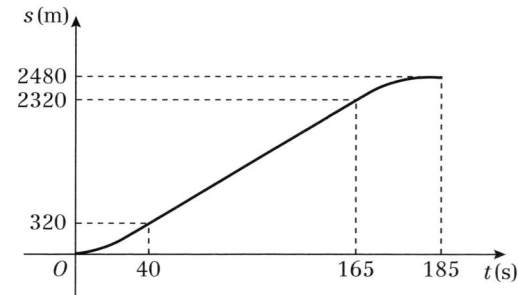

   [Graph: trapezoidal velocity-time graph rising from O to 24 at t=75, constant at 24 until time T, then decreasing to 0 over interval t_1]

   b $0.48\,\text{m s}^{-2}$ **c** $T = 250$ **d** 375 s
3. **a** 185 s   **b** 2480 m
   c $s\,(\text{m})$

   [Graph: displacement-time curve from O, reaching 320 at t=40, 2320 at t=165, and 2480 at t=185]

4. **a** $28\,\text{m s}^{-1}$ **b** 5.7 s (2 s.f.)
5. $t = 2$ and $t = 4$
6. $p = \sqrt{30}$ and $q = \sqrt{10}$
7. $66\tfrac{2}{3}\,\text{m}$
8. **a** $0.693\,\text{m s}^{-2}$ (3 s.f.)   **b** 7430 N (3 s.f.)
   c **i** Rotational forces and air resistance can be ignored.
   ii The tension is the same at both ends and its mass can be ignored.
9. **a** Ball will momentarily be at rest 25.6 m above A.
   $0^2 = u^2 + 2 \times 9.8 \times 25.6$, $u = 22.4$
   b 4.64 (3 s.f.)   **c** 6380 (3 s.f.)
   d $v\,(\text{m s}^{-1})$

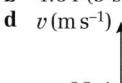

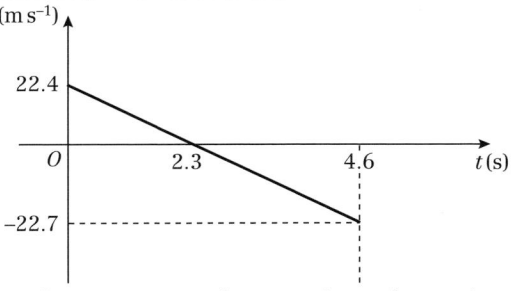

   [Graph: linear velocity-time graph from 22.4 at t=0, crossing zero at t=2.3, reaching −22.7 at t=4.6]

   e Consider air resistance due to motion under gravity.
10. **a** $4.2\,\text{m s}^{-2}$ **b** 3.4 N (2 s.f.)
    c $2.9\,\text{m s}^{-1}$ (2 s.f.)   **d** 0.69 s (2 s.f.)
    e **i** String has negligible weight.
    ii Tension in string is constant i.e. same at A and B.
11. **a** 2.9 N (2 s.f.)
    b $4.9\,\text{m s}^{-2}$
    c 0.21 s (2 s.f.)
    d Same acceleration for P and Q.
12. **a** **i** 1050 N   **ii** 390 N
    b $3\,\text{m s}^{-2}$
13. **a** 8697 N   **b** 351 N   **c** 507 N
14. **a** 63°
    b $2 + \lambda = k$ (1) and $3 + \mu = 2k$ (2)
    $2 \times (1) = (2)$ so $4 + 2\lambda = 3 + \mu$ so $2\lambda - \mu + 1 = 0$
    c 4.47 (3 s.f.)
15. **a** 17.5 (1 d.p.)
    b 66°
    c $P = 3\mathbf{i} + 12\mathbf{j}$
    $Q = 4\mathbf{i} + 4\mathbf{j}$
16. **a** $8\sqrt{2}\,\text{km}$ **b** 2 pm   **c** $(3\mathbf{i} + 6\mathbf{j})\,\text{km}$
17. **a** 5 N   **b** 7
18. $m = 50\sqrt{3} + 30$, $n = 50$
19. **a** $\sqrt{(-75)^2 + 180^2} = 195 > 150 = \sqrt{90^2 + 120^2}$
    b Boat A: 6.5 m/s; Boat B: 5 m/s; Both boats arrive at the same time – it is a tie.

Challenge
1. $t_1 = 62.2\,\text{s}$, $t_2 = 311.1\,\text{s}$, $t_3 = 46.7\,\text{s}$ (3 s.f.)
   Distance = 20.6 km (3 s.f.)
2. **a** $a = 7.4\,\text{m s}^{-2}$ **b** 39 N
   c 13 N   **d** 55 N (2 s.f.)
   e Acceleration is the same for objects connected by a taut inextensible string.

CHAPTER 5

Prior knowledge check
1. $(\mathbf{i} + 2\mathbf{j})\,\text{m s}^{-2}$
2. **a** 16.55   **b** 25.02°

Exercise 5A
1. **a** **i** 11.3 N (3 s.f.)
      ii 4.10 N (3 s.f.)
      iii $(11.3\mathbf{i} + 4.10\mathbf{j})\,\text{N}$
   b **i** 0 N
      ii −5 N
      iii $-5\mathbf{j}\,\text{N}$
   c **i** −5.14 N (3 s.f.)
      ii 6.13 N (3 s.f.)
      iii $(-5.14\mathbf{i} + 6.13\mathbf{j})\,\text{N}$
   d **i** −3.86 N (3 s.f.)
      ii −4.60 N (3 s.f.)
      iii $(-3.86\mathbf{i} - 4.60\mathbf{j})\,\text{N}$
2. **a** **i** −2 N
      ii 6.93 N (3 s.f.)
   b **i** 8.13 N (3 s.f.)
      ii 10.3 N (3 s.f.)
   c **i** $(P\cos\alpha + Q - R\sin\beta)\,\text{N}$
      ii $(P\sin\alpha - R\cos\beta)\,\text{N}$
3. **a** 39.3 N (3 s.f.) at an angle of 68.8° above the horizontal
   b 27.9 N (3 s.f.) at an angle of 16.2° above the horizontal
   c 3.01 N (3 s.f.) at an angle of 53.3° above the horizontal
4. **a** $B = 30.4\,\text{N}$, $\theta = 4.72°$
   b $B = 28.5\,\text{N}$, $\theta = 29.8°$
   c $B = 13.9\,\text{N}$, $\theta = 7.52°$

5 a $\frac{\sqrt{3}}{5}$ m s^{-2} b 48 N
6 $20\sqrt{2}$ N
7 36.3 kg (3 s.f.)
8 $2T\sin 60° + 20g = 80g$
  $T = \frac{60g}{2\sin 60°} = 20\sqrt{3}\,g$
9 $F_1 = 12\sqrt{3}$ N, $F_2 = 20$ N

Challenge
$F_1 = (6\sqrt{2} - \sqrt{6})$ N, $F_2 = (2\sqrt{3} - 2)$ N

Exercise 5B
1 a (diagram: block on incline at 20°, R normal, weight 3g)
  b 27.6 N (3 s.f.)   c 3.35 m s^{-2}
2 a (diagram: block on incline at 30°, R normal, 50 N along slope, weight 5g)
  b 42.4 N (3 s.f.)   c 5.1 m s^{-2}
3 a 3.92 N (3 s.f.)   b 5.88 m s^{-2} (3 s.f.)
4 a (diagram: block on incline at 15°, R normal, F along slope, 30 N, weight 6g)
  b 14.8 N (3 s.f.)
5 a 0.589 kg (3 s.f.)   b 4.9 m s^{-2}
6 0.296 m s^{-2} (3 s.f.)
7 15.0 N (3 s.f.)
8 R(↗): $26\cos 45° - mg\sin\alpha - 12 = m \times 1$
  $13\sqrt{2} - 12 = m + \frac{1}{2}mg$
  $m = \frac{13\sqrt{2} - 12}{1 + \frac{g}{2}} = 1.08$ kg (3 s.f.)

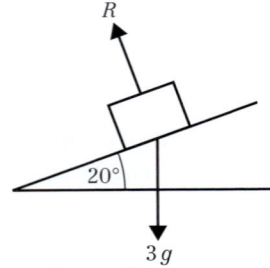

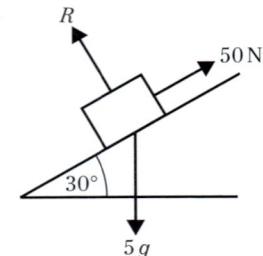

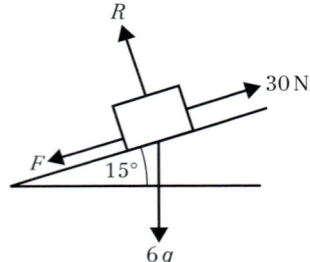

Challenge
a $mg\sin\theta = ma$ and $mg\sin(\theta + 60°) = 4ma$
  $4\sin\theta = \sin(\theta + 60°)$
  $4\sin\theta = \sin\theta\cos 60° + \cos\theta\sin 60°$
  $4\sin\theta = \frac{1}{2}\sin\theta + \frac{\sqrt{3}}{2}\cos\theta$
  $\frac{7}{2}\sin\theta = \frac{\sqrt{3}}{2}\cos\theta$
  $\tan\theta = \frac{\sqrt{3}}{7}$
b 13.9°

Exercise 5C
1 a i 3 N
    ii $F = 3$ N and body remains at rest
  b i 7 N
    ii $F = 7$ N and body remains at rest
  c i 7 N
    ii $F = 7$ N and body accelerates
    iii 1 m s^{-2}
  d i 6 N
    ii $F = 6$ N and body remains at rest
  e i 9 N
    ii $F = 9$ N and body remains at rest in limiting equilibrium
  f i 9 N
    ii $F = 9$ N and body accelerates
    iii 0.6 m s^{-2}
  g i 3 N
    ii $F = 3$ N and body remains at rest
  h i 5 N
    ii $F = 5$ N and body remains at rest in limiting equilibrium
  i i 5 N
    ii $F = 5$ N and body accelerates
    iii 0.2 m s^{-2}
  j i 6 N
    ii $F = 6$ N and body accelerates
    iii 1.22 m s^{-2} (3 s.f.)
  k i 5 N
    ii $F = 5$ N and body accelerates
    iii 3.85 m s^{-2} (3 s.f.)
  l i 12.7 N (3 s.f.)
    ii The body accelerates.
    iii 5.39 m s^{-2} (3 s.f.)
2 a $R = 88$ N, $\mu = 0.083$ (2 s.f.)
  b $R = 80.679$ N, $\mu = 0.062$ (2 s.f.)
  c $R = 118$ N, $\mu = 0.13$ (2 s.f.)
3 0.242 (3 s.f.)
4 0.778 N (3 s.f.)
5 56.1 N (3 s.f.)
6 16.5 N (3 s.f.)
7 a Use $v = u + at$ to find $a = -\frac{2}{3}$ m s^{-2}
    R(→): $-\mu mg = -\frac{2}{3}m$
    $\mu = \frac{2}{3g}$
  b The coefficient of friction remains unchanged. The air resistance has no effect on the coefficient of friction, which is dependent on the properties of the wheels and the rails.
8 1.96 m s^{-2}

Challenge
R(↗): $mg\sin\alpha - \mu mg\cos\alpha = ma$
$g\sin\alpha - \mu g\cos\alpha = a$

Chapter review 5
1 a 32.0 N (3 s.f.)
  b 0.5 m s^{-2}
2 $F_1 = 27.8$ N, $F_2 = 24.2$ N (3 s.f.)

3 a

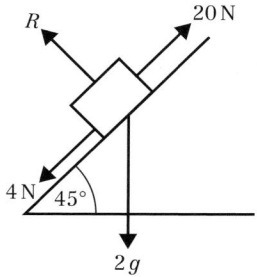

[Diagram: block on 45° incline with forces R (perpendicular), 20 N (up along slope), 4 N (down along slope), and 2g (downward)]

b 13.9 N (3 s.f.)
c Res (↗): $16 - 2g \sin 45° = 2a$
$a = \frac{16 - 2g \sin 45°}{2} = 1.1 \text{ m s}^{-2}$ (2 s.f)

4 2.06 m s^{-2} (3 s.f.)

5 R(→): $F = 150 \cos 45° + 100 \cos 30°$
$= \frac{150\sqrt{2}}{2} + \frac{100\sqrt{3}}{2}$
$= 25(3\sqrt{2} + 2\sqrt{3})$ N

6 $\mu = \frac{5\sqrt{3}}{93}$

7 11.4 N

8 3.41 m s^{-2} (3 s.f.)

9 a 4400 N (2 s.f.) **b** 0.59 (2 s.f.)
c i e.g. The force due to air resistance will not remain constant in the subsequent motion of the car.
ii e.g. While skidding, the car is unlikely to travel in a straight line.

Challenge

$F_{MAX} = 0.2 \times 400g \cos 15° = 760$ N (2 s.f.)
Component of weight that acts down the slipway:
$400g \sin 15° = 1000$ N (2 s.f.)
1000 N > 760 N so boat will come to momentary rest then accelerate back down the slope.
The boat will take 6.9 seconds to reach the water.

CHAPTER 6

Prior knowledge check

1 $2\sqrt{17}$ N at 14° above **i**
2 a 5.5 m s^{-1} **b** 6 m
3 3.2 N

Exercise 6A

1 30 m s^{-1} **2** 2.5 m s^{-1} **3** 3 m s^{-1}
4 6.5 m s^{-1} **5** 2.59 N s (2 d.p.)

Exercise 6B

1 4 m s^{-1}
2 $\frac{20}{9}$ m s^{-1}
3 4.5 m s^{-1}
4 a $\frac{8}{3}$ m s^{-1} **b** $\frac{8}{3}$ N s
5 a 1 m s^{-1} and direction unchanged
b 15 N s
6 10
7 a $\frac{2u}{3}$; direction reversed **b** $8mu$
8 Larger 8 m s^{-1} and smaller 4 m s^{-1}
9 a 3 **b** $\frac{9mu}{2}$
10 a 3 m s^{-1} **b** 4.5
11 a 4 m s^{-1} in same direction
b 3 m s^{-1} in opposite direction
12 a 3 m s^{-1} **b** 6 kg

Challenge

P: $I = \frac{9mu_1}{4}$ Q: $I = \frac{3mu_2}{2}$

$\frac{9mu_1}{4} = \frac{3mu_2}{2}$ gives $u_1 = \frac{2u_2}{3}$

Chapter review 6

1 a $\frac{1}{2}u = v$, direction reversed **b** $6mu$
2 a 14 m s^{-1} **b** $\frac{35}{3}$ m s^{-1} **c** 0.75 m (2 s.f.)
d e.g. The pile driver is likely to bounce slightly so the particles will not coalesce; the pile driver is much heavier than the pile so the particles will behave as if they coalesce.
3 a 2000 **b** 36 m
4 a 1.75 m s^{-1} **b** 0.45 N s
5 a 2.5 m s^{-1} **b** 15 000 N s
6 a 0.7 m s^{-1} **b** unchanged **c** 8.25 N s
7 $\frac{4}{5}$
8 a 7.5 m s^{-1} **b** 11 000 (2 s.f.)
c R could be modelled as varying with speed.

CHAPTER 7

Prior knowledge check

1 0.278 (3 s.f.)
2 10.1 N (3 s.f.) at an angle of 32.8° above the dashed line

Exercise 7A

1 a i $Q - 5\cos 30° = 0$ **ii** $P - 5\sin 30° = 0$
iii $Q = 4.33$ N $P = 2.5$ N
b i $P\cos\theta + 8\sin 40° - 7\cos 35° = 0$
ii $P\sin\theta + 7\sin 35° - 8\cos 40° = 0$
iii $\theta = 74.4°$ (allow 74.3°) $P = 2.20$ N (allow 2.19)
c i $9 - P\cos 30° = 0$
ii $Q + P\sin 30° - 8 = 0$
iii $Q = 2.80$ N $P = 10.4$ N
d i $Q\cos 60° + 6\cos 45° - P = 0$
ii $Q\sin 60° - 6\sin 45° = 0$
iii $Q = 4.90$ N $P = 6.69$ N
e i $6\cos 45° - 2\cos 60° - P\sin\theta = 0$
ii $6\sin 45° + 2\sin 60° - P\cos\theta - 4 = 0$
iii $\theta = 58.7°$ $P = 3.80$ N
f i $9\cos 40° + 3 - P\cos\theta - 8\sin 20° = 0$
ii $P\sin\theta + 9\sin 40° - 8\cos 20° = 0$
iii $\theta = 13.6°$ $P = 7.36$ N

2 a i [Triangle diagram: right triangle with 60° at top, 30° at bottom, sides P and $4\sqrt{3}$, labeled Q]
ii $Q = 4$ N, $P = 8$ N

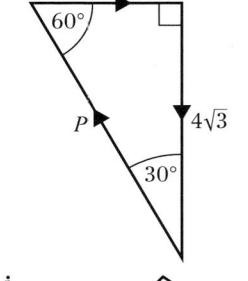

b i [Triangle diagram with sides 4 N, P N, 7 N; angles 45° and θ°]

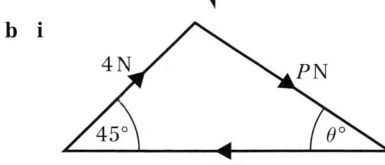

ii $\theta = 34.1°$, $P = 5.04$ N

3 a $P = 4.33\,\text{N}, Q = 2.5\,\text{N}$ **b** $P = 7.07\,\text{N}, Q = 7.07\,\text{N}$
  c $P = 4.73\,\text{N}, Q = 4.20\,\text{N}$ **d** $P = 3.00\,\text{N}, Q = 0.657\,\text{N}$
  e $P = 9.24\,\text{N}, Q = 4.62\,\text{N}$

Challenge

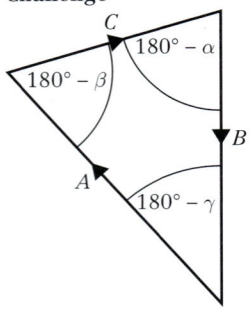

$$\frac{A}{\sin(180° - \alpha)} = \frac{B}{\sin(180° - \beta)} = \frac{C}{\sin(180° - \gamma)}$$

$$\frac{A}{\sin\alpha} = \frac{B}{\sin\beta} = \frac{C}{\sin\gamma}$$

Exercise 7B
1. $35\,\text{N}$ (2 s.f.)
2. **a** $20\,\text{N}$ **b** 1.77
3. **a** $33.7°$ **b** 14.4
4. $30\,\text{N}$ and $43\,\text{N}$ (2 s.f.)
5. **a** $5.46\,\text{N}$ **b** $0.76\,\text{kg}$
  c Assumption that there is no friction between the string and the bead.
6. **a** $1.46\,\text{N}$ **b** $55\,\text{g}$
7. **a** 2.6 **b** $4.4\,\text{N}$
8. **a** $F = 19.6m, R = 9.8m$ **b** $F' = 17m$ (2 s.f.), $R' = 0$
9. $13.9\,\text{N}$
10. $39.2\,\text{N}$
11. **a** $15.7\,\text{N}$ (3 s.f.)
  b $37.2\,\text{N}$ (3 s.f.)
  c Assumption that there is no friction between the string and the pulley.
12. $R = 0.40\,\text{N}$ (2 s.f.)

Exercise 7C
1. 0.446
2. 0.123
3. **a** $1.5\,\text{N}$ **b** Not limiting
4. **a** $40\,\text{kg}$
  b The assumption is that the crate and books may be modelled as a particle.
5. **a** $11.9\,\text{N}$ **b** $6.40\,\text{N}$
6. 0.601
7. **a** $13.3\,\text{N}$ **b** $X = 10.7\,\text{N}$
8. **a** $22.7\,\text{N}$
  b $9.97\,\text{N}$ down the plane
  c $\mu \geq 0.439$
9. **a** $X = 44.8$ **b** $R = 51.3\,\text{N}$
10. $T = 102$ (3 s.f.)
11. $2.75 \leq T \leq 3.87\,\text{N}$
12. 0.758
13. **a** 0.75 **b** $67.2\,\text{N}$

Chapter review 7
1. **a** $32.3°$ (3 s.f.)   **b** $16.3\,\text{N}$ (3 s.f.)
2. **a** $18.0°$ (3 s.f.)   **b** $43.3\,\text{N}$ (3 s.f.)
3. $T_1 = 1062\,\text{N}, T_2 = 1013\,\text{N}$
4. **a** $12.25\,\text{N}$ **b** $46.6\,\text{N}$ (3 s.f.)
  c F will be smaller

5. **a** $R(\uparrow): T\cos 20° = 2g + T\cos 70°$
$$T = \frac{2g}{\cos 20° - \cos 70°}$$
$$= 33\,\text{N (2 s.f.)}$$
  b $42\,\text{N}$ (2 s.f.)
6. Proof
7. Proof
8. Proof
9. Proof

CHAPTER 8
Prior knowledge check
1. **a** $13.1\,\text{cm}$ **b** $12.0\,\text{cm}$
2. **a** $10.8\,\text{N}$ **b** $21.6\,\text{N}$ **c** $7.8\,\text{N}$ (2 s.f.)

Exercise 8A
1. **a** $6\,\text{Nm}$ clockwise   **b** $10.5\,\text{Nm}$ clockwise
  c $13\,\text{Nm}$ anticlockwise   **d** $0\,\text{Nm}$
2. **a** $10\,\text{Nm}$ anticlockwise   **b** $30.5\,\text{Nm}$ anticlockwise
  c $13.3\,\text{Nm}$ clockwise   **d** $33.8\,\text{Nm}$ anticlockwise
3. **a** **i** $313.6\,\text{Nm}$ clockwise
    ii $156.8\,\text{Nm}$ anticlockwise
  b Sign is a particle.
4. **a** $0\,\text{Nm}$ **b** $0\,\text{Nm}$
  c $36\,\text{Nm}$ anticlockwise   **d** $36\,\text{Nm}$ anticlockwise
5. $2.5\,\text{N}$

Exercise 8B
1. **a** $5\,\text{Nm}$ anticlockwise   **b** $13\,\text{Nm}$ clockwise
  c $19\,\text{Nm}$ anticlockwise   **d** $11\,\text{Nm}$ anticlockwise
  e $4\,\text{Nm}$ clockwise   **f** $7\,\text{Nm}$ anticlockwise
2. **a** $16\,\text{Nm}$ clockwise   **b** $1\,\text{Nm}$ anticlockwise
  c $10\,\text{Nm}$ clockwise   **d** $7\,\text{Nm}$ clockwise
  e $0.5\,\text{Nm}$ anticlockwise   **f** $9.59\,\text{Nm}$ anticlockwise
3. $6\,\text{m}$
4. 1.6

Exercise 8C
1. **a** $10\,\text{N}, 10\,\text{N}$ **b** $15\,\text{N}, 5\,\text{N}$
  c $12\,\text{N}, 8\,\text{N}$ **d** $12.6\,\text{N}, 7.4\,\text{N}$
2. **a** $7.5, 17.5$ **b** $30, 35$ **c** $245, 2\frac{2}{3}$
3. $0.5\,\text{m}$
4. $59\,\text{N}$
5. $31\,\text{cm}$ from the broomhead
6. **a** $16.25\,\text{N}, 13.75\,\text{N}$ **b** $3.2\,\text{m}$
7. **a** $784\,\text{N}$ **b** $0.625\,\text{m}$
8. **a** $122.5\,\text{N}$ **b** $1.17\,\text{m}$
9. **a** $\frac{9}{2}T_C = 4W + 8 \times 30$
$\frac{9}{2}T_C = 4W + 240$
$9T_C = 8W + 480$
$T_C = \frac{8}{9}W + \frac{160}{3}$
  b $T_A = \frac{W}{9} - \frac{70}{3}$ **c** $750\,\text{N}$

Challenge
$3\,\text{kg}, 5\,\text{kg}, 1\,\text{kg}, 2\,\text{kg}, 4\,\text{kg}$ (from left to right)

Exercise 8D
1. $R_A = 2.4\,\text{N}, R_B = 3.6\,\text{N}$
2. **a** $10g\,\text{N}$ **b** $3.5\,\text{m}$ from A
3. $\frac{1}{3}\,\text{m}$ from A

ANSWERS

4 **a** 29.4 N, 118 N   **b** 2.11 m
5 **a** 160 N   **b** 2.77 m
6 **a** 3 m
   b Centre of mass lies at the midpoint of the seesaw.
   c 2 m toward Sophia.
7   $R_C = 5R_D$
   $R_C + R_D = 80 + W$
   $R_D = \frac{80 + W}{6}$
   Taking moments about A: $6R_C + 20R_D = 80 \times 10 + xW$
   $50R_D = 800 + xW$
   $25W - 3xW = 400$
   $W = \frac{400}{25 - 3x}$

Exercise 8E
1   5
2   $\frac{2}{3}$ m
3   2.05 m
4 **a** $C = 15$ N, $D = 5$ N   **b** $2 \times 12 \neq 20 \times 0.5$
   c $2.14\,\text{m} \leq x \leq 4.78\,\text{m}$
5   2.5 m
6 **a** Taking moments about N:
   $mg \times ON = \frac{3}{4}mg \times 2a$ so $ON = \frac{3}{2}a$
   b $\frac{23}{20}mg$ N
7   40 N

Chapter review 8
1 **a** 105 N   **b** 140 N
   c 1.03 m to the right of D
2 **a** $(1 \times 150) + W(x - 1) = 1.5\left(\frac{150 + W}{2}\right)$
   $150 + Wx - W = 112.5 + 0.75W$
   $37.5 = 1.75W - Wx$
   $150 = 7W - Wx$
   $W = \frac{150}{7 - 4x}$
   b $0 \leq x < \frac{7}{4}$
3 **a** 40 g   **b** $x = \frac{1}{2}$
   c i The weight acts at the centre of the plank.
     ii The plank remains straight.
     iii The man's weight acts at a single point.
4 **a** $2.5 \times 100 = 3.5W + 150(3.5 - x)$
   $250 = 3.5W + 525 - 150x$
   $150x = 3.5W + 275$
   $300x = 7W + 550$
   b $W = 790 - 300x$ **c** $x = 2.53, W = 30$
5 **a** 200 N   **b** 21 cm
6 **a** 36 kg   **b** 2.2 m
7 **a** 19.6 N   **b** 5
8   $\frac{2}{3}$ m
9 **a** 125 N   **b** 1.8 m
10 **a** $\frac{10\,000}{x}$ **b** $500\,\text{kg} \leq M \leq 2000\,\text{kg}$
   c This model has the crane only able to lift weights of 500 kg at full extension, not very practical.

Challenge
1 **a** 69.1 N   **b** 163 N

Review exercise 2
1   19.8 N (3 s.f.)
2   R(↘): $F\cos 30° - 2g\sin 45° = 4$
   $\frac{\sqrt{3}}{2}F = 4 + \frac{2g\sqrt{2}}{2} \Rightarrow F = \frac{2}{\sqrt{3}}(4 + \sqrt{2}g)$ N (as required)
3 **a** 144 767 N (to the nearest whole number)
   b 0.10 (2 s.f.)
   c 0.74 seconds (2 s.f.)

   d Will not remain at rest. $F_{\text{MAX}} = 15\,000$ N and component of weight down slope = 26 000 N which is greater.
4   6.3 N s
5 **a** 16 m
   b Air resistance would result in a greater deceleration.
6 **a** 2.25 m s^{-1}, direction unchanged
   b 1.5 N s
7 **a** 2.4 m s^{-1} **b** Direction reversed
   c 3000 kg
8 **a** A: 2.2 m s^{-1}; B: 3 m s^{-1}
   b $mu = 1.5$ N s, $mv = 1.1$ N s
   $mu - mv = 0.4$ N s
   c 1.6 N s
9   3 m s^{-1}
10 **a** 3.6 kg   **b** 18 N s
11 **a** R(↑): $T\cos 30° = g + T\cos 60°$
   $T = \frac{2g}{\sqrt{3} - 1}$ (as required)
   b 36.6 N (3 s.f.)
   c There are no frictional forces acting on the bead.
12 **a** Resolving perpendicular to the slope:
   $R = 500g\cos\alpha$ and $\cos\alpha = \frac{24}{25}$
   So $R = 480g$ (as required)
   b $68g$ N
13   0.25
14   Proof
15 **a** 0.5 g   **b** 35 m
16 **a** $\frac{g(15 - \sqrt{3})}{30}$ m s^{-2} **b** $\frac{30 + 14\sqrt{3}}{15(1 + \sqrt{3})} \approx 1.32$ m (3 s.f.)
17   Proof

Challenge
1   Taking moments about B, in limiting equilibrium:
   $0.4mgk = 0.4k \times 100 + 1.2Fk$
   $12F = 4mg - 400$, so $F = \frac{1}{3}(mg - 100)$ N
   So in order for m to be lifted $F > \frac{1}{3}(mg - 100)$ N
2   Proof
3   3 m s^{-1}

Exam-style practice
1 **a** 8 seconds   **b** 0.3 m s^{-1}
2 **a i** the train is accelerating
     ii the train is moving with constant velocity
     iii the train is decelerating
   b 4.5 hours
   c 240 km
3 **a** 2.25 m   **b** $\frac{35}{3}g$ and $\frac{25}{3}g$
4 **a** 1/5 g m s^{-2}, 12/5 g N   **b** Proof
5 **a** Proof
   b 1188 N
   c constant acceleration/constant friction
   d 25.5 m (3 s.f.)
6 **a** 7.46 km h^{-1} (3 s.f.)   **b** 4.72 hours
7 **a** Proof
   b Proof
   c 0.25 g m s^{-2} down the slope

INDEX

A
acceleration 10–38, 64–8
- connected particles 71–2
- constant 10–38
- displacement-time graphs 11–13
- due to gravity 4, 29, 64
- equation of motion 64–71
- on inclined planes 94–7
- kinematics formulae 19–29
- modelling assumptions 4
- pulley systems 75–7
- units 6, 40
- as a vector 51, 68–71
- velocity-time graphs 13–16
- vertical motion under gravity 29–35

acceleration-time graphs 17–19
air resistance 4, 7
answers to exercises 158–68
assumptions, mathematical models 4–5
average speed 11–12
average velocity 11–12

B
bead model 4, 121, 123
bearings 40–2, 62, 68
buoyancy 7

C
centre of mass 4, 139–42, 143
coefficient of friction 98
- calculating 100, 102, 126–7
collisions 108–13
column vectors, notation 63
compression 6
connected particles 71–4
- pulleys 75–9
conservation of momentum 108–13
constant acceleration 10–38
coplanar forces 120, 135

D
deceleration 13, 20–1, 26–7
- acceleration-time graphs 17–18
- velocity-time graphs 14–16
directed line segment 42–5
displacement 11–13
- kinematics formulae 19–29
- units 6, 40
- vectors 40–2
- and velocity vectors 49–50
- velocity-time graphs 13–16

displacement-time graphs 11–13
distance travelled 11–12, 40–2
- kinematics formulae 19–29
- and velocity vectors 49–50
- velocity-time graphs 13–16

drag 7
dynamics 58–82
- connected particles 71–9
- force diagrams 59–61
- forces and acceleration 64–8
- forces as vectors 62–4, 68–71
- motion in two dimensions 68–71
- Newton's laws 59–61, 64–74, 107, 108
- pulley systems 75–9

E
equation of motion 64–71
equilibrium 59, 62
- limiting 98, 100, 125–30
- modelling objects in 121–5
- and moments 136–9
- static 117–25
- triangle of forces 92, 93, 117–18

exam practice paper 153–5

F
force diagrams 6–8, 94–7
- dynamics 59–61
- statics 121–5

forces 6–8, 59–68, 88–104
- and acceleration 64–8
- component in direction of motion 89
- equation of motion 64–71
- equilibrium 92, 93, 117–25
- friction 6, 98–103, 125–30
- impulse of 106–8
- moments 132–48
- resolving 59–61, 89–97, 117–20
- as vectors 51, 62–4, 68–71
- *see also* resultant force

friction 6, 98–103, 125–30

G
glossary 156–7
gravity 4, 29–35, 64

H
horizontal surfaces 59, 89, 91
- forces and acceleration 65–8, 71–4

forces and friction 98–102, 125–8
pulley systems 76–7, 79

I

i and **j** notation 45–9, 62–4
impulse 105–15
    particle collisions 108–13
impulse-momentum principle 106–7, 108
inclined planes
    resolving forces 94–7, 118, 120
    rough 100, 102, 126–30
    smooth 94–7, 121–2, 124–5
inextensible strings 71–9, 121–3
    modelling assumptions 4, 5

K

kinematics formulae 19–29
    using with vectors 51–3
    vertical motion under gravity 29–35

L

lamina 4, 134, 136
Lami's Theorem 120
lift 6
light object model 4, 5, 135–6
limiting equilibrium 98, 100, 125–30
limiting value 98
line segment 42–5

M

mass 4, 5, 6, 40
    and acceleration 64–8
    centre of 4, 139–42, 143
    equation of motion 64–71
    and momentum 106–13
mathematical models 1–9
    assumptions 4–5
    constructing 2–3
    force diagrams 6–8, 59–61, 94–7, 121–5
    objects on inclined planes 94–7
    quantities and units 6, 7
    statics 121–5
    validity 2–4
moments 132–48
    and centres of mass 139–42, 143
    definition 133
    equilibrium 136–9
    lamina 134, 136
    light rod 135–6
    non-uniform rod 139–42, 143, 144
    point of tilting 142–4
    resultant moments 135–6
    uniform rod 136–9, 142, 143–4

momentum 105–15
    conservation of 108–13
    impulse-momentum principle 106–7, 108
    in one dimension 106–8
motion 10–38, 58–82, 105–15
    equation of 64–71
    Newton's laws 59–61, 64–74, 107, 108
    in two dimensions 68–71
    vertical under gravity 29–35

N

newtons 6
Newton's first law of motion 59–61
Newton's second law of motion 64–71
Newton's third law of motion 72–4, 107, 108
non-uniform rod 139–42
    on point of tilting 143, 144
normal reaction 6, 59, 65–6, 91, 122
    and friction 98, 99, 101–2
    on inclined planes 94–5, 96

P

parallel vectors 42, 43, 48–9
parallelogram law of vector addition 43
particles 58–82, 116–31
    acceleration 19–35, 51
    collisions 108–13
    connected 71–9
    in equilibrium 57, 62, 92
    forces acting on 51, 59–61, 91–4
    on inclined planes 95–7, 100
    in limiting equilibrium 98, 125–30
    mass 64–5, 106
    model 4, 5
    momentum 106–13
    position vectors 50–1
    in static equilibrium 117–20, 123–5
    velocity 14–15, 49–53, 106
    vertical motion under gravity 29–35
peg model 4
pivots, point of tilting 142–4
position vectors 50–1
projectiles 29, 31–5
pulleys 4, 75–9
    static 121–2, 124–5

R

resolving forces 59–61, 89–97
    inclined planes 94–7, 118, 120
    static particles 117–20
resultant, definition 42
resultant force 59–61, 68–71
    and acceleration 64–8

static particles 117
vector addition 62–4, 91–2
vector forces 62–4, 68–71
resultant moments 135–6
retardation *see* deceleration
review exercises 83–7, 149–52
constant acceleration 35–8, 83–5, 87
dynamics 79–82, 84–7
forces and friction 103–4, 149
mathematical modelling 8–9, 84–5
moments 144–7, 152
momentum and impulse 113–14, 149–50
statics 130–1, 150–2
vectors 54–6, 85–6
rigid bodies 4, 117, 132–48
rods 4, 6, 135–44
connected particles 71–4
non-uniform 139–42, 143, 144
on point of tilting 142–4
uniform 136–9, 142, 143–4
rotational forces 4
see also moments
rough surfaces 59, 98–103, 125–30
inclined planes 100, 102, 126–7, 128–30
mathematical models 4, 6

S
scalar quantities 40, 106
SI units 6, 21
smooth surfaces 4, 98
inclined plane 94–7, 121–2, 124–5
speed 6, 11–12, 40
calculate using vectors 49–50, 51
and momentum 106–13
speed-time graphs 14–16
statics 116–31
and friction 125–30
mathematical models 121–5
resolving forces 117–20
static equilibrium 117–25
subtracting vectors 43, 47
suvat formulae *see* kinematics formulae

T
tension 6, 71–9, 121–3
thrust 6, 7
tilting 142–4
time 6, 11–19, 40, 50–3
triangle of forces 92, 117–18, 119
triangle law of vector addition 42–3, 91–2

U
uniform acceleration *see* constant acceleration
uniform body model 4
uniform rod 136–9
on point of tilting 142, 143–4
unit vectors 45–9, 51, 62
units 6, 21, 106, 133

V
validity of mathematical models 2–4
vector equation of motion 68–71
vector quantities 40, 106
vectors 39–57
acceleration 51, 68–71
adding 42–5, 47, 48–9, 91–2
directed line segments 42–5
displacement 40–2
equation of motion 68–71
force 51, 62–4, 68–71
i and **j** notation 45–9, 62–4
magnitude 47
multiplication by a scalar 43
parallel 42, 43, 48–9
position 50–1
solving velocity and time problems 50–3
subtracting 43, 47
unit vectors 45–9
velocity 49–53
zero vector 57
velocity 6, 13–16, 19
acceleration-time graphs 17–19
displacement-time graphs 11–13
kinematics formulae 19–29
and momentum 106–13
particle collisions 108–13
of a particle as a vector 49–50
solving problems with vectors 50–3
velocity vectors 49–53
velocity-time graphs 13–16, 19
vertical motion 29–35

W
weight 6–7, 40, 59, 64
and friction 98–100
statics problems 121–5
uniform/non-uniform rods 136–44
wire model 4

Z
zero vector 57